New Frontiers in
Technological Literacy

NEW FRONTIERS IN
TECHNOLOGICAL LITERACY

BREAKING WITH THE PAST

Edited by John R. Dakers

Cover photo by John R. Dakers

First published in 2014 by PALGRAVE MACMILLAN® in the United States—a division of St. Martin's Press LLC, 175 Fifth Avenue, New York, NY 10010.

Where this book is distributed in the UK, Europe and the rest of the world, this is by Palgrave Macmillan, a division of Macmillan Publishers Limited, registered in England, company number 785998, of Houndmills, Basingstoke, Hampshire RG21 6XS.

Palgrave Macmillan is the global academic imprint of the above companies and has companies and representatives throughout the world.

Palgrave® and Macmillan® are registered trademarks in the United States, the United Kingdom, Europe and other countries.

ISBN: 978-1-137–38632-8

Library of Congress Cataloging-in-Publication Data

New frontiers in technological literacy : breaking with the past / edited by John R. Dakers.
    pages cm
  Includes bibliographical references and index.
  ISBN 978-1-137-39474-3 (hardback)
  ISBN 978-1-137-38632-8 (paperback)
  1. Technological literacy.   I. Dakers, John R.

  T65.3.N448 2014
  600—dc23   2014000299

A catalogue record of the book is available from the British Library.

Design by Scribe Inc.

First edition: July 2014

10 9 8 7 6 5 4 3 2 1

For my lovely lady

Cover image description:

Machinic Assemblage
A term coined by Deleuze and Guattari referring to the way different connections form different forms of expression. Using For example:

Cart + Flowers + Locale + Bearing = Ornament
Cart + Donkey + Merchant + Children + Beach = Seaside Pleasure Trip
Cart + Horse + Hawker + Salvage = Rag and Bone man
Cart + Horse + Farmer + Corn = Harvest

Contents

FOREWORD

ADVANCING AND QUESTIONING LITERACY

Carl Mitcham

Defining Technological Literacy (2006) was a good book. *New Frontiers in Technological Literacy* advances that project in insightful and beneficial ways.

The previous volume neatly brought together 18 original articles by a mixture of established and younger scholars from five countries (Australia, Canada, Netherlands, the United Kingdom, and the United States). The present collection adds ten new scholars from five more countries (China, Denmark, Germany, New Zealand, and Spain). The consequence is to extend, in the words of Albert Borgmann's foreword to the original book, an effort to understand "the moral and cultural quality of the world" being constructed globally through technology.

The literacy at issue here is not simply one enabling us to be more successful within the techno-global system. Instead, it aims to deepen critical engagement from the overlapping perspectives of Chinese experience (Wang), democracy (Williams), education (Alonso and Barlex), gender (Kirk), literacy itself (Dakers and Petrina), medicine (Samerski), sustainability (Elshof), work (Wallace and Hasse), and youth (Watson).

In *Electric Rhetoric* (1999), a book that has exercised considerable influence on rethinking literacy, Kathleen Welch called for us to develop a consciousness of ourselves as techno-social beings. Such consciousness gives life to an inside-out paradox. How do we understand ourselves as persons at once substantively informed by the technologies with which we live and yet able critically to assess these technologies? How can we step outside the technologies we are in? Is criticism ever able to do more than simply extend technology?

Take the case of education. Ivan Illich argued in the 1970s that professionalized schooling had become counterproductive insofar as it had become a way to tie students ever more tightly into the technosocial world. Having concluded that his own alternative educational institution, the Centro Intercultural de Documentación (CIDOC), was itself in danger of becoming part of the system, he closed it and became an itinerant scholar for the last three decades of his life.

During those last years, Illich turned to what he called an archeology of modern certainties, literacy being one of them. In *ABC: The Alphabetization of the Popular Mind* (1988) and *In the Vineyard of the Text* (1993), he sought to disclose how the techniques of reading transform the readers. In like manner, Illich argued, modern technologies emerge from and then reinforce a distinctive ethos, the recognition of which is best appreciated by investigations into the moral environments of previous techniques.

Such an approach led Illich to become critical of certainties on both the left and the right: notions of "environmental responsibility" and what he saw as the new ideology of "life." Calls for environmental responsibility were, he argued, often just another excuse for advancing technological management of the world; the Christian pro-life movement, while proclaiming to criticize science, defined human life in terms of a molecular-biological genesis that escapes any direct experience.

What was at work in history was a counterproductivity writ large that Illich described with a Latin phrase: *corruptio optimique est pessima*, or the corruption of the best is the worst. To quote Shakespeare,

> For fairest things grow foulest by foul deeds;
> Lilies that fester, smell far worse than weeds.

We must be cognizant of the danger that in promoting technological literacy we do not too uncritically do more for technology than for literacy. Many of the chapters in this collection take precisely such a critical stance—and we should be thankful to John Dakers for shepherding such works to publication.

ACKNOWLEDGMENTS

I would like to give particular thanks to Wendy Dow for her continued help, advice, and encouragement. I am especially grateful to her for her continued support.

I would also like to thank Burke Gerstenschlager, Scarlet Neath, Mara Berkoff, and Sara Nathan from Palgrave Macmillan for their patience, encouragement, and advice in assisting me with the production of this volume.

I am deeply indebted to Carl Mitcham, who has been a great support and inspiration to me in the production of this volume. Carl not only has offered guidance but has actively assisted in the production of this book. I am also indebted to Don Ihde, who continually offers me encouragement and inspiration. I also wish to acknowledge a huge debt to Marc de Vries, who has, for many years, always supported and encouraged me to develop my thinking in this area.

Finally, I would like to thank the various authors for their help, advice, support, and contributions.

INTRODUCTION

BREAKING WITH THE PAST

John R. Dakers

In this companion edition to my first book, *Defining Technological Literacy: Towards an Epistemological Framework*, now published as a significantly revised second edition, I have brought together several new thinkers, each of whom makes an important contribution to the concept of technological literacy. The title of this book, *New Frontiers in Technological Literacy*, suggests a move into hitherto uncharted territory in terms of thinking related to technological literacy. The subtitle *Breaking with the Past* indicates a move away from traditional arguments about how technological literacy might be incorporated into extant technology education classroom practices. I make no apology for my views on this; it cannot be done, it has not been done, it will not be done. The incorporation of the concept of technological literacy into the prevailing model of technology education is akin to realizing the truly admirable plea made by C.P. Snow in his seminal book *The Two Cultures*, first published in 1959 (Snow, 2001). Snow postulates that society is split into two cultures; one related to the sciences and one related to the humanities. This dichotomy was, for Snow, resolvable. However, the passing of time has served to indicate otherwise: science and the humanities continue to act as two distinct cultures. I believe that technology education had its genesis, and continues to align itself strongly with, the scientific culture of positivism. Technology education openly associates itself with engineers and engineering. As Feenberg elucidates, "Once the object is stabilized, the engineer has the last word on its nature, and the humanist interpreter is out of luck. This is the view of most engineers and managers; they readily grasp the concept of 'goal' but they have no place for 'meaning'" (Feenberg, 1995: 9).

The concept of technological literacy, on the other hand—a relatively new paradigm in technology education—sits more comfortably within the humanities. The growing body of research and philosophical discourse relating to technological literacy tends to situate it around issues specific to the social aspects of technology: the impact that technology has on human beings and the human-technology interface.

This reenactment of Snow's dilemma as realized in the form of technology education is, in my view, no more resolvable now that it was then. If we want to promote a form of technological literacy, especially in human terms, we need to consider a new way to think about technology. Carl Mitcham elucidates a "critical dimension" in the foreword to this book, one that we must come to recognize will change the way we consider, design, and interface with technology. In so doing, we must correspondingly critique how this will change the environment we share and change our way of communicating with that environment, as well as with each other. The future, in this sense, is subject to multiple possible technologically textured trajectories, all of which can continue to be shaped by the few in power or by initiating change, by becoming critical, by becoming technologically literate.

Becoming technologically literate is, in my view, too important to be sidelined and marginalized in education, in terms of both today and the future. The contributions that follow in this book offer some insightful arguments relating to the concept of technological literacy, all discussed from a diverse range of overlapping perspectives.

The book first sets out to examine the very concept of technological literacy. In my own chapter, "Technological Literacy as a Creative Process of Becoming Other," I offer a new pedagogy: a pedagogy that considers a completely new way of enabling the development of technological literacy. I call this pedagogy "speculative multidimensional time line thinking." Drawing heavily on the work of Deleuze and Guattari, I attempt to make a case for becoming technologically literate—not "being" technologically literate, or finally becoming technology literate, as these scenarios denote a telos, a final end state: that of being technologically literate. I argue that due to the complexity of the technologically textured world we inhabit, a world that is emergent and in a constant state of change, the concept of technological literacy can only ever be expressed in terms of an ongoing process.

In this sense, the concept of a technological world considered as being essential, fixed, and enduring, a technological world that we tend to adopt as received, must be flawed. Instead, I suggest an alternative way to consider the technological world in terms of change: one that considers technology historically as well as having potentially variable alternative historical pathways that regard technology as having multiple capacities, hidden as well as known. By so doing, we can perceive the technological world in a nonlinear way, full of possible pasts, presents, and futures. In this way, a form of speculative technological literacy can be conceived that will enable us to consider the world from different perspectives.

Stephen Petrina, in his chapter "Postliterate Machineries," deconstructs the concept of literacy from both a historical perspective and a technoscientific conception of literacy. Also drawing heavily on the works of Deleuze and Guattari, he links technology to literacy by way of machineries and then to the concept of postliterate machineries. In so doing, he enquires whether, in a postliterate world, we still need literacies. Indeed, he questions the very idea of the concept of postliteracy. But as his enquiry expands, it becomes evident that he regards this question as too simplistic. The world today is extremely complex, and the future cannot be known. It becomes clear for Petrina that we need to go beyond the grounded, stable definitions of literacy that prevail today. We need to reconceptualize the concept of literacy technologies as well as technological literacies. To do this, we need to be creative and to experiment by exploring unknown pathways leading to unknown potentials in terms of literacy.

Molly Watson is a 15-year-old school student who offers a chapter titled "Technology and Technology Education: Perspectives from a Young Person." There are many academic books devoted to various aspects of research into education and technology, virtually all of which are written by adult academics. These books consider many divergent perspectives on education and technology. Many result from empirical investigations that do listen to young peoples' voices. This chapter is different in that I asked Watson, someone I know, if she would be interested in writing a chapter based on her perspective of technology. She was very keen to engage with this project. Sometime later, she produced about two thousand words on how she, and some of her friends, constituted technology. I then asked her to write some additional material relating to what she thought technology education was. It becomes very apparent in her chapter that, at least from her perspective, there is a significant, if not an overwhelming, dislocation between her perceptions of technology and the school subjects she thinks represent that genre.

Watson offers a balanced view on technology. She clearly sees benefits and also highlights some serious concerns relating to technology. She offers a useful insight into the social aspects related to technology. Technology is not some neutral artifact that exists separately from humans, "because when you think about it, what is technology without the people that use it?" In terms of technology education, she indicates a view that suggests a bias toward a pedagogy of prescription, a pedagogy predicated on the transmission of already given "facts," a pedagogy of essentialism: "There is too much focus in the current technology curriculum on following orders. Most young people I know do not want this. And perhaps the reason why many teenagers dislike school is because they are not given sufficient opportunities to express themselves, and they are not presented with things that they feel are relevant. And if something is relevant, they are not always told why it is relevant." This chapter endorses much of the thinking expressed in the other chapters in this book.

By considering "Technological Literacy and Digital Democracy: A Relationship Grounded in Technology Education," P. John Williams looks at affects related to globalization. He suggests that technology education during colonial times was antidemocratic and had a bias toward the Western conception espoused by the colonizers. This meant that any form of technological literacy taught was equally biased and often inappropriate for indigenous populations.

Moreover, he goes on to highlight that, as a result of globalization, a universal curriculum has emerged that is ignorant of local values. Williams uses this dichotomy to explore and critique the delivery of technology in several countries around the world. He then considers the impact that new and emerging digital technologies are having on technology and technology education, particularly in relation to issues relating to democracy. He concludes that technological literacy must aspire toward the formation of "fully developed digital citizens who are able to interpret, critique, navigate, and shape the landscape of virtual democracy."

Mary Kirk raises several important issues relating to gender in technology. The title of her chapter is very revealing: "Reenvisioning Our Knowledge Tradition: From Gender Blind to GenderAware." She asserts that the way that we have learned to think about technology is related to our assumptions about gender. Technology is not gender neutral, as we tend to think, but is gender blind.

Kirk outlines the stereotypical gender-assigned assumptions we hold and demonstrates the way in which these assumptions form strong boundaries that influence our conceptions about technology. This dualistic way of categorization has, over the course of history, offered a male-centric view of technology and science. However, Kirk demonstrates the contributions to the fields of technology and science that have been made by what she refers to as the "invisible women of science and technology." She concludes that "gender blindness" in technoscientific thought continues to remain stubbornly encoded in our knowledge tradition. She asserts that a consideration of the history of women's contributions to technology and scientific disciplines, which should form part of becoming technologically literate, may help to change the imbalance that continues to prevail.

Leo Elshof offers an incisive critique of the sustainability crisis that is threatening the planet in his chapter "Eco-Technological Literacy for Resiliency." He highlights issues that indicate the rise of indifference in relation to the way we treat our planet, the consequences of which indicate that we are actually living beyond the biocapacity of the planet.

In an attempt to elucidate an understanding of this crisis, he deconstructs the relationship between consumption, neoliberal capitalism, and globalization. As a way forward, Elshof suggests that technological literacy should involve aspects relating to eco-technological literacy. This, he suggests, will offer students a way to develop a more mindful connection with our technological creations. However, he sees this not as a return to the old craft-orientated systems of education that still prevail as the dominant orthodoxy in technology education today but rather suggests that technological literacy and ecotechnological literacy should be about innovation, collaboration, and connectivity.

The debates concerning technological literacy tend be dominated by Western culture. In her chapter, Nan Wang offers "A Chinese Perspective on Technological Literacy." In a fascinating account, Wang illustrates that China "manifests a more positive appreciation of technology than is often the case in the developed world." This, she asserts, is largely due to technology being considered as a way to both reduce the burden of human labor and increase human productivity. However, the rise of technological development in China is a relatively new phenomenon, and as such, the negative aspects regarding technological development that are debated in Western cultures are only beginning to emerge in China.

In terms of literacy, Wang outlines the significant cultural influences that form around Chinese thinking about technology. These influences offer an insight into the dualistic way a culture might be affected by technological development, as well as how that culture will itself affect technological development.

David Barlex offers a critique of the policies, politics, and delivery of design and technology education in England and Wales in his chapter "Enabling Both Reflection and Action: A Challenge Facing Technology Education." There will be many common attributes that are recognizable in many other countries around the world to the perspective offered by Barlex. Technology education for Barlex is more than learning the various techniques associated with technology. It has the power to develop the imagination to create new worlds. This can only be made possible if the pedagogical framework for technology education allows for both action and reflection. This, Barlex argues, constitutes a form of technological literacy.

Barlex explores several initiatives that he has been involved with over the years: initiatives that enable the kind of pedagogy discussed. He goes on to describe the research and development that occurred as a result of these various initiatives, as well as the developments that were adopted in practice. However, he also highlights the political dimensions that serve to enable, or indeed to block, the ongoing development of projects that can only be considered as progressive. Moreover, and perhaps disturbingly, Barlex implies that the subject known as design and technology in England and Wales—a subject that has been at the forefront of technology around the world—appears uncertain.

Andoni Alonso introduces the concept of cyberliteracy in his chapter "From Cybereducation to Cyberactivism: Can Cyberliteracy Transform the Public Sphere?" Not unlike Williams earlier, Alonso questions whether Information and Communication Technology can liberate access to education, thus democratizing education in ways never before achievable or even conceivable. He illustrates the impact that electronic media have on education and goes on to problematize several of the unforeseen impacts that have emerged as a result of this new form of learning. Areas such as cyberscience and cyberactivism are considered in some detail as ways to illustrate the threat that "the colonizing interests of private profit generating commerce" are having on the liberation of education and therefore democracy.

The chapter by Jamie Wallace and Cathrine Hasse moves the concept of technological literacy into work-based environments. In "Situating Technological Literacy in the Workplace," Wallace and Hasse argue that any understanding of what constitutes technological literacy remains tethered to formal systems of education. Their chapter investigates the concept of technological literacy in terms of workplace environments. This moves the debate from an essentially theoretical perspective to one that is situated—one that is very much context dependent. Hospitals and schools form the basis of their empirically driven enquiry.

Significantly, they report that technology, as made available in the workplace, is not something fixed that workers can simply learn and use or that enables greater efficiency in the workplace. Rather, they argue that technology "exists in relation to the unfolding consequences of processes and ways of thinking and organizing mutually constituted between social and technological worlds." In this respect, it becomes evident that the use of technologies in the workplace exceeds the primary functions ascribed to them. It becomes clear that they also have economic, political, and ethical dimensions and that this has implications for workers. Wallace and Hasse argue that technological literacy enables workers to develop "the capacity for learning from everyday entanglements within the constant reconfigurations of both practice and technology without losing sight of the motive for practice itself."

Finally, Silja Samerski in her chapter titled "Genetic Literacy: Scientific Input as a Precondition for Personal Judgment?" provides a revealing discussion centered on the concept of genetics, highlighting how experts appear to purposely avoid any opportunity for a wider dialogue with nonexperts. Genetic literacy thus appears to be supercilious to nonexperts. The author then questions such superciliousness and makes a demand for a more enlightened engagement regarding communications in the technoscientific discourse.

These chapters combine to offer multiple perspectives on the concept of technological literacy. It is not necessary to read them in any given order. The chapters offer alternative perspectives that are situated in some context, whether in the actual technologically textured world of school, work, or culture; the virtual world of cyberspace; the sociopolitical world of gender or the environment; the philosophical world; or the world as experienced by a young person. Whatever the context, these overlapping perspectives merge, not just as presented in this book, but with the philosophical, political, socioeconomic, and many other perspectives that have been expressed in the past. Moreover, these perspectives will continue to merge with the readers of this book, who will agree with, disagree with, and hopefully reconstitute the thoughts expressed in new and interesting ways for the future. I welcome this ongoing process.

REFERENCES

Feenberg, A. (1995). Subversive rationalization: Technology, power, and democracy. In Andrew Feenberg &Alistair Hannay (Eds.), *Technology and the politics of knowledge* (pp. 3–22). Bloomington: Indiana University Press.

Snow, C.P. (2001 [1959]). *The two cultures.* London, UK: Cambridge University Press.

CONCEPTUALIZING
TECHNOLOGICAL LITERACY

CHAPTER 1

TECHNOLOGICAL LITERACY AS A CREATIVE PROCESS OF BECOMING OTHER

John R. Dakers

INTRODUCTION

Samuel Butler offers a somewhat dystopic view of Victorian society in his now famous satire, *Erewhon*, published in 1872. In it, he tells the story of a fictitious country where the strange inhabitants have formed a society that forbids the use of modern technology (only old, established, and nonthreatening technology is allowed). This society actively suppresses any expression of originality, experimentation, or creativity. It further contends that individual scholarship can only ever be made manifest in a curriculum that develops proficiency in the study of what is referred to as "unreason and hypothetics." Indeed, a venerable Erewhonian professor of worldly wisdom states that "it is not our business to help students to think for themselves. Surely that is the very last thing which one who wishes them well should encourage them to do. Our duty is to ensure that they shall think as we do, or at any rate, as we hold" (189).

Despite the superb efforts of many of my colleagues from around the world to challenge this perspective, I believe that technology education, in its present incarnation, for the most part, fits rather neatly into Erewhonian culture. In this chapter, I want to explore a completely new alternative for the delivery of technology education; an alternative that will focus exclusively on the development of technological literacy. In so doing, I hope to promote (provoke) some discussion about alternative pathways, as well as alternative potentialities that we might consider, or indeed discover.

Drawing primarily on the work of Gilles Deleuze, a French poststructuralist philosopher who incidentally was influenced by Butler's *Erewhon*, I shall explore the concept of "becoming technologically literate." This reflection differs from the many other important questions frequently asked about technology education. Rather than explore technology education as something homogeneous, I will instead approach the subject from a heterogeneous perspective. In so doing, I will deconstruct the notion of *technology education* as a distinct and objective subject domain that studies the given technologies and techniques that apply to the known world that we, as human beings, have constructed for ourselves. Instead, I will reconstruct the world as one made up of multiple dimensions and perspectives, one distinct from static objective realities. This will represent a progressive and radical departure from the world of education as elucidated in cultures similar to *Erewhon*.

ESSENTIALISM AND CLASSIFICATION

The world that we occupy today continues to be shaped by the philosophy of Plato and Aristotle. Despite being teacher and student, both philosophers had significantly different perspectives on the way that we humans perceive the world. However, they have both had an enduring impact on the world of philosophy and on the way that we human beings continue to make sense of the world.

Plato developed the concept of ideal forms. These ideal forms represent the intrinsic nature or indispensable quality of something that determines its character: a property of something without which it would simply not exist; a thing's essence. Significantly, for Plato, these forms were not subject to alteration; they were already given, already established perfect manifestations of all things hidden somewhere in the dark recesses of the mind. In order to know something or do something, one has to have a conception of its ideal form, its ultimate state of perfection, its essence: "Plato believed that the essence of a thing is the form in which it participates" (Korsgaard, 1996, p. 2). In other words, it is the ideal form that constitutes perfection. This applies equally to organic things, nonorganic things, and metaphysical things like a poem. Human beings, for example, tend to be inclined, depending on circumstances and motivational factors, to strive toward perfection. We continue to be, alas, perennially disappointed in this endeavor. However, this continual striving toward perfection constituted the ethical question that was, for Plato, of primary importance: "how should one live?" or "what constitutes a good life?" At the summit of this order is the form of the *good* to which all human beings must strive toward (June, 2011, p. 91). For Plato, these ideal forms can never actually be achieved or sensorially experienced by human beings. They can only be known, and human beings know these ideal forms intrinsically: they are genetic forms of knowledge that have been predetermined for us (Korsgaard, 1996; June, 2011). By way of illustration, consider a potter or a carpenter turning a bowl. They want to produce a bowl that is perfectly cylindrical in form, but clearly, no human being is capable of creating the ideal form of a perfect circle, so the bowl produced will constitute a copy of the ideal cylindrical form, an ideal circle. The closer the potter or carpenter is able to get to creating a perfect circular form, the closer to perfection he or she is able to get. This Platonic form of the good is a process or an activity that has orientated human beings toward leading their lives such that their endeavors would ultimately manifest in *arete*—that is, to strive toward excellence, toward being good at what they are, to become virtuous. However, virtue expressed in terms of *arete* can only ever be judged in terms of its relation to the general ideal form of the good, just as the potter's cylindrical form can only ever be judged in terms of how close it resembles an ideal circle; we are never likely to achieve it, but it serves as a guiding principle. Examinations follow this line of reasoning to some extent: the closer to (re) presenting the preestablished ideal form, the higher the grade achieved; the further away, the lower the grade achieved. Examinations, particularly technology education examinations, leave no room for alternative perspectives.

Plato's ontological arrangement in the form of the good is, however, trenchantly rejected by his student Aristotle, who asserts "the fundamental existence of sensible particulars against that of general entities. The species and genera that classify individuals exist, but they are only secondary in the ontological scheme" (Lawson-Tancred, 1995, pp. 406–407). In other words, Aristotle rejects the idea that particular definitions must ultimately originate from or be determined by some general, superior, and hierarchical ideal form: "If one asks what something is, which is to ask for its definition, then the only sort of informative answer that can be given [for Aristotle] is one that refers to its species. And if one asks what that is, then one must refer to the subgenus in which it is most immediately included, and so on up to the highest genus to which it can belong, which is that general category of substance" (Lawson-Tancred, 1995, p. 407).

Aristotle's categories, then, are phenomenal and realized, not in the Platonic ideal sense, but as something substantial. The Aristotelian categories, writ large, are genus, species, and individual. This system of classification has formed the bedrock for the way we have come to identify and understand the world to this day. Significantly, it is not confined to biological systems, but it informs a system of classification that covers the entire spectrum of categorization that prevails in all disciplines, including those associated with technoscientific studies and technology education. Typical forms of assessment in technology education include questions that relate to categories: name different types of wood (hardwood, softwood, and human-made boards), for example.

"Aristotle's *Categories* is a singularly important work of philosophy . . . [that has] engaged the attention of such diverse philosophers as Plotinus, Porphyry, Aquinas, Descartes, Spinoza, Leibniz, Locke, Berkeley, Hume, Kant, Hegel, Brentano and Heidegger (to mention just a few), who have variously embraced, defended, modified or rejected its central contentions. All, in their different ways, have thought it necessary to come to terms with features of Aristotle's categorical scheme" (Studtmann, 2007). Aristotle, in his treatise, attempts to enumerate the most general kinds of categories (genera), subdivide them into workable entities (species), and then finally reduce them to

their essential qualities (individuals). These categories enable us to differentiate between objects given in the world and so make sense of the world. A cat is
not a dog, nor is a tree a mountain, for example. A cat has an essential quality
that is categorically different from the essential qualities representing a dog.

Both Plato and Aristotle rely on essential qualities in order for their theses
to make sense. As such, their worlds can only ever be perceived of from an
anthropocentric perspective. It is human beings, as subjects, who discover
and (re)present the world from their own perspective along the lines explicated by Plato and Aristotle.

CONVENTIONAL THOUGHT ON THE
CONCEPT OF A *WHOLE*

I have previously asserted that there is no such thing as a car (Dakers, 2014).
In conventional thinking, a car is thought of as an objective *whole* entity,
one that has some essential enduring quality expressed in terms of the parts
that combine to give *it* or *them* an enduring identifiable trait: "The essence
of a thing is that which explains its *identity*, that is, those fundamental traits
without which an object would not be what it is. If such an essence is shared
by many objects, then possession of a common essence would also explain
the fact that these objects *resemble* each other and, indeed that they form a
distinct *natural kind* of things" (Delanda, 2002, p. 9; italics in original).

In other words, there are defining characteristics that explain an object
and so differentiate it from other objects. In so doing, these characteristics combine to form part of a classifiable, identifiable species, or "wholes,"
that will all resemble each other (trees, humans, cows, or cars, for example).
Thus, in conventional thinking, a whole is generally considered to be something that is logically deducible from general principles. For example, it is
universally accepted that the general principle of combining a variety of specific entities—wheels, windscreens, seats, engines, and so on—in a specific
way will form a quantitative whole that we have come to know as being
a car. Moreover, a whole—in this case a car—will also have certain essential qualities that are expressed in terms of its parts. A car is essentially the
combination of a number of recognizable and established components that,
when combined, are identifiable as belonging to our conception of what
constitutes a car. If, for example, only two wheels were to form part of the
ensemble, we would reconceptualize the whole as being a bicycle or a motor-
bike. A less complex example of a whole is water. The combination of two
specific atoms in a specific way, one of oxygen and two of hydrogen, will
form a quantitative whole that we have come to understand as being water,
or H_2O. A whole in conventional terms, then, is an entity—whether material
or immaterial (a car, water, a human, or thoughts expressed in the form of a
poem, for example)—that comprises the sum of its parts, is identifiable (or at
least has the capacity to be identified), and is, significantly, subject to dissolu-
tion or complete loss of identity should a component part be removed. If we
remove a component from a car—its engine, for example, which is in itself a

whole—we would completely change the constitution of the car. It would no longer be recognized as being a car, given that a car, thought in conventional terms, requires an engine as part of its established identity. In this respect, a whole can only be considered as being a whole, providing it is capable of being given a definitive classification. An entity that cannot be fully classified (a car without an engine, for example), can never be fully known and so cannot be considered to be whole in the terms discussed. A human being who does not fit the ideal form (Plato), or does not conform to the essential classification that constitutes a human being (Aristotle), cannot be considered to be definable as a human being. While they may resemble human beings, they need extended forms of classification in order to differentiate them from the essential form to which they are related (e.g., handicapped, schizophrenic, slow, hyperactive).

Given, then, that conventional thinking considers that a whole can only be understood in terms of it having a specific and definitive identity, that identity must imply a boundary of some sort, one that serves to distinguish one whole from another. A "normal" human being having two legs, for example, is distinguished from a human being who has only one leg. A classificatory boundary is established in order to differentiate the two. Another obvious example is a map. A map of Europe defines the boundaries that distinguish one country from another. The whole of Germany is identified as something distinct from the whole of Italy. Each country is, in turn, made up of identifiable areas that are also distinguished by towns and cities and so forth. The distinguishing areas or parts that together constitute the whole of Germany distinguish it from Italy, which forms another whole country made up of different parts. Moreover, other distinct parts help formulate the essential quality that constitutes Germany and make it distinct: language, architecture, gastronomy, and culture, to name but a few. Germany was, however, in the past, two distinct wholes: East Germany and West Germany. They were considered to be two separate entities then: two identifiable, enduring, and bounded countries each having different political and cultural identities. However, by redefining the boundaries in all respects, they have come together to form a new whole known as Germany. Change has occurred. The enduring properties of two formerly distinct and identifiable wholes have altered, and boundaries have been redefined. This puts into question the enduring properties of wholes as predicated in conventional thought. The boundaries defining two identifiable entities have become porous, enabling the formation of a novel entity, one having new identifiable characteristics.

Nevertheless, no matter what whole it is that we are dealing with as human beings, we continue to define wholes, or entities, as separate, identifiable wholes that can only ever be analyzed in terms of their parts. Thus we define wholes, ultimately, in terms of their essential properties. Our entire world is classified in this way especially as a result of scientific classification predicated on an Aristotelian system of classification. There are, thus, very clear *scientific* identities, and so boundaries, imposed on the concept of wholes that tend to externalize, objectify, and separate things into clearly identifiable, enduring, stable components ranging from atomic structure to the properties of the universe. This is how we human beings continue to make sense of the world.

Wholes can be technological, metaphysical, or naturally occurring entities. A car or a computer may be considered to be a technological whole, the complete works of Shakespeare as a metaphysical whole (presented in some actual form like a book or a play), while a tree or a human being may be considered as a natural whole. Whether technological, metaphysical, or natural conventional thinking requires that wholes are able to be reduced to their essential qualities. However, Deleuze (1988) and later Deleuze and Guattari (1987) offered an alternative way to consider the world by introducing and developing the concept of assemblages.

THE CONCEPT OF ASSEMBLAGES

The philosophy of Deleuze is complex, and any full explanation of his oeuvre and associated terminology are considerably beyond the scope of this chapter. One very important concept in Deleuzian thought, however, is that of the *assemblage*. Some of the explanations offered for an assemblage in this book may be considered as overlapping with another important Deleuzian concept known as *multiplicity*. While the two concepts differ in degree, they can be interpreted as having some common attributes depending on the context in which they are set. In order to simplify my interpretation of an assemblage, I have elected to use only the term *assemblage* throughout this chapter.

Following the mathematics developed originally by Gauss and later expanded by Riemann, a new form of mathematics was developed that completely changed aspects of the mathematics developed by Descartes. This is in combination with the philosophy developed by Bergson, Deleuze, and Guattari offer a calculus of change: nonenduring wholes that exist without having been given some external set of coordinates, as determined by some external hierarchy, such as to render an established universally accepted identity for any given whole. They call this concept an assemblage.

An assemblage, for Deleuze and Guattari, is not constrained by some essential quality and has no enduring universal identity. It can and does have boundaries, but these boundaries are porous and subject to constant variation (like the unification of Germany). Universal classifications therefore no longer apply. Concepts such as country, school, technology education, student, state, society, and culture, for example, are assemblages—concepts that have no enduring properties but rather possess properties that are subject to constant states of change, redefinition, and reconstitution.

Assemblages, for Deleuze, are differentiated into two types: quantitative and qualitative. The former is a relatively straightforward concept: quantitative assemblages are "actual, objective, and extensive. [They] are represented in space, posses an identity, and differ in degree from one another" (Tampio, 2010, p. 912). Qualitative assemblages, on the other hand, cannot be counted; they are "virtual, subjective and intensive [and] are experienced in lived time" (p. 912). Both forms of assemblage, Deleuze suggests, are intertwined; they coexist and interpenetrate: "Deleuze takes the idea that [technology, for example] is composed of different [assemblages] that form a kind of patchwork or ensemble without becoming a totality or whole" (Roffe, 2005, p. 176).

Consider a house, for example. In conventional thinking, a house is an objective whole that can be identified as a place where people live. A house has actual properties that can be quantified, like walls and floors and a roof. It has rooms to live in, to cook in, and to bathe in. A house has windows and doors and rainwater pipes. The problem with this description is, however, that it does not offer a clear, logical, mathematical, or scientific definition for a house, one that can reduce the concept of *house* to some essential quality. No description can. This is because there is no essential quality that can represent a house. As soon as we try to offer one, there is always something else, something more, something beyond, something we had not considered possible before. We simply do not know what a house is capable of becoming in the future. We do know, however, what the concept of a house has been in historical and cultural terms: a cave, an igloo, a tepee, a yurt, a bungalow, an apartment, a cardboard box, to name but a few. But none of these offer us any universal, enduring identity that offers us some quality that represents the specific essence of a house. It only offers some general conception related to history and culture.

While a house can be considered a quantitative assemblage in that it is actual and identifiable as a house, albeit in a contingent sense, it is also a qualitative assemblage. A house is more than the sum of its quantifiable parts; it is a form of expression, a form of intensity that awakens emotions and feelings. A house can offer protection from the elements; it may be considered to be a sanctuary, an enclosed space offering privacy, warmth, and security. A house may, for some, appear to express the wealth of its owners, or the power of its inhabitants (as in the case of the White House), or again it may express aesthetic beauty for others. A house, which is an assemblage, is capable of both affecting and being affected.

Put another way, there is no such thing as a house, at least not in the conventional sense. There is only becoming house, not becoming *a* house, as that would presuppose some ideal final form. For Deleuze, "things and states are products of becoming. The human subject, for example, ought not to be conceived as a stable, rational individual, experiencing changes but remaining principally, the same person: rather, for Deleuze, one's self must be conceived as a constantly changing assemblage of forces, an epiphenomenon arising from chance confluences of language, organisms, societies, expectations, laws and so on" (Stagoll, 2005, p. 21).

The Anatomy of an Assemblage

An assemblage for Deleuze is not an assembly in any conventional sense. It is not a group or a collection of things brought together for some specific purpose. It is not some determinable whole that constitutes an unchanging essence, an established thing that we can establish an identity for. An assemblage is, rather, an "emergent whole," a process. Assemblages can, for example, be an individual human, an individual community, an individual organization, an individual city, or even an individual hammer. Assemblages can be "individual atoms and molecules, individual cells and organisms, individual species, and ecosystems. All these different assemblages are born at a particular time, have a life, and then die" (DeLanda, 2011, p. 185). This being the case, it becomes evident that while an assemblage may appear to endure (like a mountain or a major city, for example), it is nevertheless subject to constant change: sometimes imperceptible, as when the wind erodes parts of a mountain; sometimes dramatic, as when a demolished building modifies the skyline of a city. Assemblages constitute movement rather than stable states, and these movements are considered by both Deleuze and the English philosopher Alfred North Whitehead to be events. For Deleuze and Whitehead, everything is an event: "The world, [Whitehead] says, is made of events, and nothing but events: happenings rather than things, verbs rather that nouns, processes rather than substances" (Shaviro, 2009, p. 17). An assemblage is thus an event in a constant state of emergence, one that is constituted by the interactions among its parts: "This implies that the identity of an assemblage is always contingent and is not guaranteed by the existence of a necessary set of properties constituting an unchanging essence" (DeLanda, 2011, p. 185). There is no such thing as a house. A house is an event, one that is constituted by the interactions between its parts—human beings, extensions, new windows, burst drain pipes, floor coverings, furniture, lighting, and so on. Moreover, no two assemblages are ever the same. No two houses are ever the same; they are always in a state of constant emergence: different inhabitants, different floor coverings, different furniture, and so on. Consider, then, two technology education classes (or any classes for that matter) being taught one after the other. They each have the following properties: they are from the same school, are from the same year group, have the same number of students, have the same teacher teaching the same lesson in the same classroom for the same duration of time. Any teacher who has done this, and many will have, will agree that despite the similarities, the dynamics of the two lessons will nevertheless be different. The two lessons, or events, can be considered to be assemblages, wholes that emerge in slightly (or perhaps significantly) different ways.

Assemblages are not unique, novel events that happen as if by magic. While they can be prearranged up to a point, the resultant outcome of their existence can never be known in advance. All assemblages, while pointing to the future, are affected and influenced by the past. They have a history that influences the coming together of their various properties, capacities, and tendencies: "[T]he identity of an assemblage should always be conceived as the product of a historical process, the process that brought its components together for the first time as well as the process that maintains its integrity through a regular interaction among its parts" (DeLanda, 2011, p. 185).

The two technology education scenarios given here, if actual, would each have been the product of a historic process. Every component in each assemblage—the teacher, the students, the theoretical underpinnings of the lesson, the pedagogy employed, and the tools and materials used—is the product of a historic process, a process that has resulted in these various components being brought together at a particular time, thus influencing it. Influencing it, but unable to determine its outcome. This is why a teacher's set of lesson plans can only ever be indicative and cannot prescribe actual outcomes. Assemblages as actuated in the present will be affected by the past, but not prescribed by it. Equally, potential assemblages that may become actualized, the outcome of which cannot be fully known, will serve to affect the future. As Marx's famous maxim indicates, human beings "make their own history, but they do not make it just as they please: they do not make it under circumstances directly encountered, given and transmitted from the past" (Marx, 1968, p. 97). An assemblage, then, does not happen in some predictable, linear fashion where all factors are known in advance. Lesson plans come to mind once again. Despite the best and most careful planning, no matter how much a teacher knows her subject, no matter how much she knows about the learners in her class, lessons rarely turn out exactly as expected, and only rarely do they result in any predetermined outcome (at the end of the lesson the learners will be able to . . .). There is always more than we could ever have anticipated. While we cannot predict the future with any certainty, we can affect it. The future is always influenced by the past.

The two classroom assemblages, therefore, while similar, are actually different emergent wholes "because these emergent wholes are defined not only by their properties [and their histories] but also by their tendencies and capacities" (DeLanda, 2011, p. 185). Each of the classroom assemblages we have considered has certain properties that are required in order to give it an identity. One such property that a classroom assemblage needs, as we have suggested, is students. All classroom assemblages require students in order to be identified as a class. In an ideal world, in the Platonic sense, two essential qualities that students are required to aspire toward are to be attentive and well behaved. Unfortunately, many students have a tendency to be inattentive and misbehave. Tendencies in an assemblage thus "make any list of essential properties look falsely permanent" (p. 186). Capacities, on the other hand, unlike tendencies, remain potentialities until they become actual. I will attempt to illustrate capacity by using the same example given by DeLanda: looking at an artifact in the form of a knife.

While one actual property of a knife is sharpness, it has a tendency to get blunt. This tendency can be overcome by sharpening the knife. Sharpening is an actual activity that is required to be done to the knife in order to counteract its tendency to get blunt. Tendencies are, thus, events that happen independently and are actual. Unlike tendencies, capacities are not actual but potential. Given that the knife is kept sharp, it also has the capacity to cut, for example. While the knife may never actually be used to cut, it will always have the capacity to cut as long as the knife is kept sharp. In order, however, for that capacity to cut to be realized and not simply to remain a potential requires that something other than the knife exists, something that is not the knife but is something that can be cut by the knife. Capacities, then, can only ever be actualized in conjunction with something that they can combine with in order to realize some given or indeed unknown capacity. To use a knife requires that the knife be in a state of sharpness. To use a knife to cut requires not only a sharp knife but also something else that is able to be cut by the knife. For example, to attempt to use a knife to cut cast iron is not likely to result in a satisfactory outcome: "The reason for this is that the [knife's] capacity to affect [to cut] is contingent on the existence of other things, cuttable things, that have the capacity to be affected by it" (DeLanda, 2011, p. 4).

To summarize, the component parts of an assemblage have histories. These histories are the result of past assemblages or events that are, in turn, the result of past assemblages or events and so on. Any change that will be constituted as a result of the formation of an assemblage can never be fully known in advance. This means that there are multiple potential pathways that an emerging assemblage may take or might have taken in the past. In other words, while only one emergent pathway will have been actualized in the past, or will be actualized in the future, other potential futures have existed in the past or may exist in the future. The scientists who developed the atomic bomb in the Second World War could have refused to do so. They appear to have wished that they did refuse to so do in retrospect. This constitutes an alternative emergent potential pathway that might have been taken, a pathway that would have presented another possible future. In order to understand better why that alternative pathway was not pursued, we would need to examine the various components that came together to form the assemblage or event that we have come to know as a human-made form of destruction. Human-designed forms of destruction have existed long before the atom bomb and have continued to proliferate since. They are constantly upgraded and changed. They are in a state of becoming other. Technologies, in whatever form, are assemblages that inevitably include human beings. They constitute emergent events that are always in a state of becoming other than that which they were previously. They have properties that are subject to change and alteration. They have embedded tendencies that become evident over time and they have capacities that have been purposefully designed as well as unknown capacities that may be revealed over time (e.g., the use of airplanes and buses as weapons). In order to have a better understanding of the technologically textured world that we have occupied, do occupy, and will occupy in the future, we need not to become technologically literate, but we need, like the world, to keep becoming other, to keep becoming technologically literate.

BECOMING TECHNOLOGICALLY LITERATE

A fictional conversation between Albert Einstein and Marilyn Monroe is depicted in the movie *Insignificance* (1985), directed by Nicolas Roeg. It offers an example of a potential assemblage, an experimental and creative way to look at what might potentially have been actualized if these two characters from history were to have a conversation about the moon:

Einstein: If I were to tell you that the moon were made out of cheese—would you believe that?
Marilyn: Of course not.
Einstein: But now if I tell you it's made out of sand?
Marilyn: Maybe . . .
Einstein: If I tell you, "I know for sure"?
Marilyn: Then I would believe you.
Einstein: So you know the moon is made of sand?
Marilyn: Yes.
Einstein: But it isn't.
Marilyn: I only said I knew because you said you knew!
Einstein: I lied. Knowledge isn't truth. It's just mindless agreement. You agree with me, I agree with someone else, we all have knowledge. We haven't come any closer to the truth of the moon. You can never understand anything by agreeing, by making definitions. Only by turning over the possibilities. That's called thinking. If I say I know, I stop thinking. As long as I keep thinking, I come to understand. That way, I might approach some truth.
Marilyn: That's the best conversation I ever had!

An actualized assemblage, the elements of which include you, the reader, this chapter, and this book have been changed by considering together a virtual or potential assemblage, one in which Einstein has a conversation with Monroe, as realized in the form of text (and a movie). If this affects you in any way, change has occurred, you have been affected and may go on, as part of another assemblage, to affect something else. You may be inspired to go and buy the movie or to change your perspective about the concept of knowledge, for example. Other things will, however, continue to affect you and you will continue to affect other things. Life, in this respect, is not a linear process with a beginning, a middle, and an end. We all start anything, including life itself, from somewhere in the middle. You do not begin to read this book without first knowing how to read. You do not end reading this book by knowing everything there is to know. You never become something, some final state. You continue becoming something other: you never become old, you are rather, always becoming other, becoming older, becoming younger, becoming wiser, becoming . . . There is no beginning, there is no end. There is only becoming.

In terms of technological literacy, one never becomes technologically literate; that would presuppose reaching a definable end state, the state of being technologically literate. It also presupposes a state that an individual human being achieves or does not achieve: a binary state. One cannot be almost technologically literate in conventional thinking; one either is or is not. Furthermore, it presupposes a fixed, determinable essential quality that identifies the conditions necessary in order to achieve a state of being technologically literate. The concept of technological literacy would require it to be universally definable, certifiable, and examinable under these circumstances. I want to argue that one can only ever be in a state of becoming technologically literate and that this can only be considered in terms of assemblage theory. In order to make this possible as a form of pedagogy, I want to introduce a concept that I will call "speculative multidimensional time-line thinking."

SPECULATIVE MULTIDIMENSIONAL TIME-LINE THINKING

In 1960, Rod Taylor starred in a film adaptation of H. G. Wells's *The Time Machine*. The film won an Oscar for best special effects, one excellent effect being when the professor is seen to move forward and backward through time. While he and the time machine do not move position relative to earth, they are perceived to move both forward and backward in time. As a stationery observer, the professor witnesses an ever-changing landscape as time passes. On the way forward, buildings disappear and new ones emerge, wooded areas grow then disappear, and a mountain is seen to envelop the professor as he moves on through time. Eventually he reaches a destination that intrigues him, not one he planned for. How could he have? Interestingly, he ends up in exactly the same place he started out from relative to earth, it just so happens to be thousands of years into the future. Clearly, change is seen to have occurred albeit speeded up. Time-lapse cinematography makes this illusion possible. It also enables us to witness plants opening up before our very eyes by speeding up the process. In contrast, it lets us watch a humming bird appear to fly in slow motion. In time-lapse photography we can witness the world differently. We are able to see things we could never see before, we can see change happening that we could never observe before by either speeding up the film process or slowing it down. My interest in the time-lapse process is, however, related to the possibilities offered in cinematography, such as the example given in the movie. To observe the plant or the humming bird as described involves a technological innovation that can capture an actualized event and show it in fast or slow motion. It does not depict potentiality. Using the same technology, as part of a narrative process in a movie, however, allows us to envision the various potentialities and possibilities that may be open to us in the future, or the influences of past potential futures on the present. Nevertheless, this type of time-lapse photography is still restricted to a linear pathway, a pathway into the future or into the past. Likewise, time-lapse thinking would also be restricted to a linear pathway. In order to open up the many possible past futures that could have influenced the future, I want to incorporate the concept of assemblage theory into what I call time-line thinking. This will allow for a speculative multidimensional approach that will not only enable other possible projected technological becomings to be envisioned from any point in history but also reveal how these projections are dependent on many other associated possible potential becomings—becomings that will have served to influence the trajectory of the life of any given technological development under scrutiny.

The speculative dimension thus becomes a creative project, one that has, as a starting point, a need to first interpret the prevailing conditions that existed at the time, in order for a technological development to change from being a potential to one that becomes actualized. These conditions, such as the material objects available at the time, the extant techniques and knowledge, and the human actors all form part of an assemblage, an event.

By developing an understanding about the relationships and connections that combined in some way to form an event, one is then better able to speculate about the possible alternative pathways that might have endured had alternative conditions prevailed. This speculative process can be used to map alternative possible futures that may have been actualized had alternative potentialities become actualized. This way of thinking about technology can also be used to consider future possible pathways for technological potentialities that prevail today. Technological literacy relating to issues such as cloning, the development and impact of cyborg technologies, and issues relating to privacy can all be considered in much more depth by adopting this methodology.

This way of thinking about technology as a creative endeavor is not an entirely new concept. Aldous Huxley gave us alternative possible futures in his *Brave New World*, several of which have become actualized. Television programs such as *Star Trek* and *Fringe*, speculate on possible technologically influenced futures, many of which appear credible. Science fiction continues to become evermore considered in terms of "science fact" as contemporary authors offer speculations about the future that are grounded in credible assemblages from the past. As a young person in the sixties, I marveled at the concept of a wristwatch that was also a televisual form of communication as speculated by Chester Gould, the creator of the comic strip character Dick Tracy. Fifty years later, I now regularly communicate, via my iPad, with my granddaughter on FaceTime. Watches that have this feature are becoming actualized today.

Some Perspectives Involved in Speculative Multidimensional Time-Line Thinking

As has already been discussed, it is necessary to consider the historical aspects of any given technology before one can speculate on various possible alternative potentialities. After having investigated and thought about the historical aspects, speculative multidimensional thinking starts from the perspective of individual agency: how does an existing communications technology, for example, affect an individual and how does that individual change as a result? This will be different in degree for everyone. For example, I am very anxious about joining a social networking site for a number of reasons, the fear of my privacy being compromised being one. I have lived a lot longer that many younger members of social networking sites, who for the most part, do not appear to be so anxious about joining, and this may be one of the factors differentiating our perspectives. However, I have many friends and colleagues of a similar age to me who are members of social networking sites. There are no universal demographics at work here, just individual agency made manifest in the different ways that we are all affected by a technology, and how that causes us to affect other technologies over time and so affect other human beings. If everyone shared my perspective about social networking sites, they would probably not exist, at least not in the way they do now. My involvement, as part of the assemblage known as social networking sites, manifests in my tending to pull in the opposite direction from the architects of these sites as well as their many satisfied users. This may not always be the case. Change will happen and happen for many, hitherto unknown reasons. Speculative multidimensional time-line thinking can not only help explore and speculate about these future potential trajectories; it can also seek to explore the reasons as to why different individuals and groups either like or do not like to participate in social networking sites.

A second perspective in multidimensional time-line thinking is the concept of the middle. As discussed earlier, every technological event starts somewhere in the middle: it has a past that influences its present incarnation, which will, in turn, influence its potential to change and become actualized differently in the future. If a technology does not change over time but continues to exist, it simply becomes a repetition of what went before—something that, for whatever reason, resists change. But any resistance to change is a factor worth considering in speculative multidimensional time-line thinking because resistance to change will affect in some respect, just as change will affect in some respect. An important and significant factor in this perspective is to recognize that there are many possible future trajectories for any technological event; some may be resistant to change, some speculative, and many unknown. For example, human beings follow one of many possible future trajectories every day. If I choose to stop working on this chapter now, I give up the alternative possible trajectory of continuing to work on the chapter. The choice I make will influence the direction I take next and that will impact the future. My train of thought may be disturbed, I may have less time to complete the work, I may prefer to go for a drink, or I may choose to watch television. Indeed, every choice I follow has at least one alternative. The path I follow therefore will be different, in varying degrees, from any alternative I might otherwise have followed. This is the same for everything. An assemblage forms by chance, something emerges as a result, and change occurs. The level of change is a matter of degree. The level or degree of change, or resistance to change, can have an influence beyond the individual. It can have little or no influence on the future or it can have significant influence with global implications. This applies to past futures as well as to actual futures. In other words, given that assemblages have at least two possible futures, one actualized and one potential, they have always had at least two possible futures. Speculative multidimensional time-line thinking offers a way not just to question the genesis of a given technology but also to explore the various events that led to the technologies emergence, to the changes that occurred over its history that have led to its present incarnation.

None of these changes are the subject of a linear process. Many associated overlapping assemblages come into existence as a result of changes, which then enabled other overlapping assemblages to form. The stronger the overlap, the stronger the connections will be. Assemblages are fractal in nature.

Finally, all technologies have properties that render them identifiable at any given time. However, as has been discussed earlier, they also have capacities and tendencies. Early computer technology had a tendency to crash. Technologies also have capacities to affect, some of which have been actualized, some of which remain potentialities, and many other capacities that either are fringe capacities or are as yet unknown. Technological capacity in terms of effect is rarely given much consideration in school-based technology education today. Remember that capacities are dualistic in nature; they require something other to affect in order to actualize their potential. A knife requires something cuttable in order to cut; a cyberbully requires someone capable of being bullied in order to bully.

Thinking in this way—thinking in terms of assemblages, thinking in terms of Speculative multidimensional time-line thinking—therefore, helps reveal a multitude of alternative possibilities that any technology might have followed or might actually follow in the future.

The Need for a Paradigm Shift toward the Creation of an Avant-Garde Pedagogy

I have presented the notion of speculative multidimensional time-line thinking as a way to change our perception of the world as an enduring and stable entity, much in the same way that the special effects team in the movie made it possible to visualize change over time. But in order to utilize the concept of speculative multidimensional time-line thinking in a practical sense, we need to institute a paradigm shift in the structure of the pedagogy that is traditionally used in conventional teaching and learning scenarios. We need to move beyond a process of knowing established technological "facts" to one of unknowing. In his essay on "Not Knowing" Brown (2013) elucidates this rather well:

> Education usually culminates in an examination, which is virtually the only time when not knowing an "answer" is unacceptable. At all other times, "I don't know" is both honorable and useful, providing that it does not reflect intellectual torpor and that an attempt is made to change the situation. This admission represents the critical identification of an absence of something necessary for understanding. Moreover, students need to become comfortable with, although not accepting of, their own ignorance because, as I have already suggested, the "unknown" greatly exceeds the "known."

This has significant implications for formal assessment procedures. Speculative multidimensional time-line thinking is not about repetition. It has no interest in the conformity of ideas. It is very much a creative endeavor, an endeavor that promotes experimentation through speculation. It requires a pedagogy of change that favors the interesting over the actual; a pedagogy of heterogeneity that recognizes multiple perspectives as more important than universal narratives; a pedagogy that gives voice to young people; a pedagogy that is recognized, in itself, as being subject to constant change and reinterpretation.

Let me try to put this into some context. I will offer a detailed example of an actual historic technological scenario that will serve to illustrate a pedagogical framework for speculative multidimensional time-line thinking: the actualization of the atom bomb. However, any of the areas discussed in this volume could also be used.

By investigating the various prevailing conditions that enabled the actualization of the atom bomb, one is led to several prior historically important developments. These developments form a linear timeline and might include the following:

1898   Marie Curie discovers radioactivity
1905   Albert Einstein theorizes about the relationship between mass and energy
1925   First cloud-chamber photographs of nuclear reactions
1932   The existence of Neutrons is proven
1934   First patent application for producing a nuclear chain reaction; aka nuclear explosion
1938   German scientists demonstrate nuclear fission
1939   Einstein warns President Roosevelt about the German potential to produce a bomb
1942   Manhattan project is formed to secretly build a bomb before the Germans
1945   The invention of the atomic bomb
1945   The USA drops atomic bombs on Hiroshima and Nagasaki

(Bellis, 2013)

This timeline offers a perspective on the events that led up to the actualization of the atomic bomb. However, they only offer a partial perspective and do not engage the reader with how things might have been otherwise. By investigating any one of these events as an assemblage, one is able to consider, in much more depth, the prevailing conditions that enabled the particular development to become actualized. In 1898, for example, Marie Curie did not simply discover radioactivity. Like every human being, she had a history that offered alternative pathways. These pathways may have led her to pursue alternative lives—lives that may not have resulted in her discovering radioactivity. Others were also involved in the discovery of radioactivity either by helping her or by influencing her. For example, the discovery of x-rays played an important role in her discovery. It is possible that had things been otherwise, radioactivity may not have been discovered, but while this is an unlikely scenario, it forms the basis of a critical examination of alternative pathways possible at the time and the resultant impact these alternatives might have had on the world today. By creating possible alternative techno-logical histories, one is able to speculate on possible technological futures that offer alternative worlds. Using the same example, if one considers the assemblage known as the Manhattan Project, one can speculate a future that did not actualize an atomic bomb and the possible future worlds that might have emerged as a result.

The conditions that led up to the creation of the Manhattan Project are well known. They include issues such as history, groundbreaking science, politics, war, fear, and curiosity. These various components were brought together to form an assemblage known as the Manhattan Project, which resulted in a change that witnessed the creation of the atomic bomb. A great many factors fed into this event, and a number of alternative future potential pathways were never actualized. How might things have been different, and how might another pathway have emerged? It is now well known that a sig-nificant number of the scientists were concerned, to say the least, about the result of their invention.

> Upon witnessing the explosion, its creators had mixed reactions. Isidor Rabi felt that the equilibrium in nature had been upset as if humankind had become a threat to the world it inhabited. Robert Oppenheimer, though ecstatic about the success of the project, quoted a remembered fragment from the Bhagavad Gita. "I am become Death," he said, "the destroyer of worlds." Ken Bain-bridge, the test director, told Oppenheimer, "Now we're all sons of bitches". After viewing the results several participants signed petitions against loosing the monster they had created, but their protests fell on deaf ears. (Bellis, 2013)

What if the scientists, as one important element of this assemblage, had refused to develop the atom bomb either before or after having witnessed the test explosion? How might that alternative world have become other than the one we know today? It was undoubtedly clear to the scientists that what they were developing was a weapon of mass destruction, one that had an awesome capacity to kill and destroy on a scale never before witnessed. In order to actualize this potential, it would clearly be necessary to have a large group of human beings, settled in some form of community somewhere, in order to kill and destroy—somewhere like Hiroshima, for example. Without this component, the atomic bomb would only have a capacity to kill and destroy. This lays bare a double aspect to the development of this technology: the actualization of a bomb with a capacity to kill and destroy on a massive scale on the one hand, together with the ethical and political question of real-izing its potential. It is unlikely, in my view, that the atomic bomb would not have been actualized as part of an unfolding of technoscientific progress. It is likely, however, that alternative possible futures might have come to pass if the decision to use the bomb had been otherwise.

By considering these possible alternative futures, a learner, in the process of becoming technologically literate, is able to speculate about technology in a creative way—one that opens up other possible technologically textured

worlds that are open to alternative perspectives, alternative political influences on technological developments, and alternative ethical dimensions, all
of which are less concerned with how the technological world *ought to be*
and instead more interested in how the technological world *might have been,
might be,* or *might become* other.

In terms of learning in this new paradigmatic framework, Guattari, in his
essay on the concept of *transversality,* identifies two types of groups that can
be translated into groups of learners: independent learners and dependent
learners. The independent learner, the type necessary for this avant-garde
pedagogy, he postulates,

> Endeavours to control its own behaviour and elucidate its object, and in this
> case can produce its own tools of elucidation. [It] could [be said] of this type
> of [learner] that it hears and is heard, and that it can therefore work out its
> own system of hierarchizing structures and so become open to a world beyond
> its own immediate interests. The dependent [learner] is not capable of getting
> things into this sort of perspective the way it hierarchizes structures subject to
> its adaptation to other [learners]. One can say of the [independent learner] that
> it makes a statement—whereas the dependent [learner] only that its cause is
> heard, but no one knows where or by whom, or when. (Guattari, 1984, p. 14)

If becoming technologically literate is considered to be an important aspect
of human development, and I believe it is, teaching and learning environments
need to change. They need to stop imposing an established technology curriculum onto learners. Rather, they need to work with learners in the creation
of concepts about how technology might become other. This means that traditional pedagogies and traditional classroom practices must become open to
constant appraisal, reformation, and reappraisal. Becoming technologically literate is thus a dynamic and fluid enterprise comprising porous and overlapping
boundaries. Formal methods of assessment, by their very nature, impose strict,
impenetrable boundaries that resist change. Resistance to change equates to a
reproduction of the same. Reproduction of the same implies a requirement to
conformity or, put another way, forms a barrier to creativity. Reproduction of
the same, when considered in terms of speculative multidimensional time-line
thinking, reveals many hierarchic systems that have existed in the past, continue to exist in the present, and will continue to influence the future. Speculative multidimensional time-line thinking serves to reveal how reproduction
of the same can significantly affect a cultures development. It also offers a way
to consider how things might have been different if certain hierarchic assemblages had led to alternative possible technology education futures.

REFERENCES

Bellis, M. (2013). *History of the atomic bomb and the Manhattan Project.* Retrieved
October 5, 2013, from http://inventors.about.com/od/astartinventions/a/
atomicbomb.htm.

Brown, S. (2013). "Not knowing" as a pedagogical strategy in the sciences. *Radical Pedagogy 10*(2). Retrieved September 13, 2013, from http://www.radical pedagogy.org.

Butler, S. (1985). *Erewhon*. London, UK: Penguin Books.

Dakers, J. R. (2014). Toward a philosophy for (*a new*) technology education. In John R. Dakers (Ed.), *Defining technological literacy: Toward an epistemological framework* (2nd ed.). New York, NY: Palgrave Macmillan.

DeLanda, M. (2002). *Intensive science and virtual philosophy*. London, UK: Continuum.

DeLanda, M. (2011). *Philosophy and simulation: The emergence of synthetic reason*. London, UK: Continuum.

Deleuze, G. (1988). *Bersonism*. New York, NY: Zone Books.

Deleuze, G., & Guattari, F. (1987). *A thousand plateaus: Capitalism and schizophrenia* (Brian Massumi, Trans.). New York, NY: Continuum.

Guattari, F. (1984). *Molecular revolution: Psychiatry and politics (peregrines)*. New York, NY: Puffin.

Insignificance. (1985) Movie released in 1985. Directed by Nicolas Roag. Produced by Alexander Stuart and Jeremy Thomas. Recorded Picture Company Zenith Productions. United Kingdom.

Jun, N. (2011). Deleuze, values, and normativity. In N. Jun & D.W. Smith (Eds.), *Deleuze and ethics* (pp. 89–107). Edinburgh: Edinburgh University Press.

Korsgaard, C. M. (1996). *The sources of normativity*. Cambridge, UK: Cambridge University Press.

Lawson-Tancred, H. (1995). *Aristotle*. In A. C. Grayling (Ed.), *Philosophy: A guide through the subject* (pp. 398–439). Oxford, UK: Oxford University Press.

Marx, K. (1968). *Selected works*. New York, NY: International Publishers.

Roffe, J. (2005). *Multiplicity*. In Adrian Parr (Ed.), *The Deleuze dictionary* (pp. 176–177). New York, NY: Columbia University Press.

Shaviro, S. (2009). *Without criteria: Kant, Whitehead, Deleuze, and aesthetics*. Cambridge, MA: MIT Press.

Stagoll, C. (2005). *Becoming*. In Adrian Parr (Ed.), *The Deleuze dictionary* (pp. 21–22). New York, NY: Columbia University Press.

Studtmann, P. (2007). *Aristotle's categories. Stanford Encyclopedia of Philosophy*. Retrieved August 9, 2013, from http://plato.stanford.edu/entries/aristotle -categories/.

Tampio, N. (2010). *Multiplicity*. In M. Bevir (Ed.), *Encyclopedia of political theory* (pp. 912–913). Thousand Oaks, CA: Sage. doi:10.4135/9781412958660.n294. Retrieved September 10, 2013, from http://www.sage-ereference.com/political theory/Article_n294.html.

CHAPTER 2

POSTLITERATE MACHINERIES

Stephen Petrina

One might just as well be machinic rather than literate, in the way one would "rather be a cyborg than a goddess." Literacies are certainly "legion" but reach semantic saturation or exaggeration against an analog of mediation and machination.[1] The sheen of the "new" is worn and tarnished, yet literacies are wont to saturate, while exhaustion sets in against a failure to reduce or subject everything to literal experience. Of course, the saturation of half of the thesis is well explored and exploited, but the machinic counterpart to the literate is entirely underplayed. Whether preliterate, aliterate, literate, or postliterate, the technological is characterized by machineries. Technologies and literacies are inseparable, subjected to the service of one another, but "machineries" productively generate a wide range of pre-and postliterate practices. To simplify, literacies signify reading and writing, while machineries signify processing and designing; literacies signify acquisition and gatherings, while machineries signify diffusion and assemblages. Both have realized significant semantic expansion and basically signify the creation of meaning, although for the latter, it is more a process of machining. With no intention of negating the literate, the goal is to recognize generations and significations of machineries over time. Documenting the exhaustion of literacies, this chapter informs and elaborates our conversation about what we have, know, or can acquire with what we became or what is becoming of human-machine assemblages, diffusion, and cyborgenic machinations. Henceforth and once again, claims staked on dimensions of natural, cultural, and artificial experience are contested: is it literacies or machineries at work and play?

Given the ancient *ars memorativa* and *technologia*, medieval mnemotechnics, or the rise of *memoria technica*, data, and information in the eighteenth and nineteenth centuries, it may have appeared that literacies evolve from and rest on machineries. Historically or archetypically, this is often portrayed as the light of literacies overcoming the dark of machineries, or as the cultured mind of literacies triumphing over the cunning hand of machineries. However, a less humanistic narrative obtains: literacies and machineries are at the least conflated or analogical and cooperational. Machineries generate practices that cannot be reduced to or subjugated and subdued by the literate, providing robust clues to machinic properties of human beings.[2] This trick, a reversal of power characteristic of machineries, can lead to a rediscovery of aliterate technicalities, instrumentalities, and monstrosities.

MACHINERIES[3]

With the "organology" of Canguilhem and "mechanology" of Simondon, Deleuze and Guattari contradicted psychoanalysts' configurations of desire as a defective mechanism, such as that found in the Oedipus complex. Subsequently, through the 1970s and 1980s, they described a vast array of machines hitherto unexplored. Hence the opening section of *Anti-Oedipus*:

> What a mistake to have ever said *the* Id. Everywhere *it* is machines—real ones, not figurative ones: machines driving other machines, machines being driven by other machines, with all the necessary couplings and connections . . . Producing-machines, desiring-machines everywhere, schizophrenic machines, all of species life: the self and the non-self, outside and inside, no longer have any meaning whatsoever . . . Thus we cannot agree with Victor Tausk when he regards the paranoiac machine as a mere projection of "a person's own body" and the genital organs . . . in and of itself the paranoiac machine is merely an avatar of the desiring-machines.

In "Psychoanalysis and Ethnology," a section published in English two years prior to *Anti-Oedipus*, they define *desiring-machines* as "the microphysics the unconscious, the elements of the micro-unconscious. But, as such, they never exist independently of the historical molar aggregates . . . Symbols and fetishes are manifestations of desiring-machines." As linguists debated the genesis of language and literacies, so did Deleuze and Guattari explore the psychic and historical genesis of machineries. They begin by clearing up confusion: "The unconscious constructs machines which are machines of desire, whose use and functioning schizo-analysis discovers in their immanent relationship with social machines. That unconscious says nothing, it machines." A machinic element, such as "a graphic system independent of the voice . . . not aligned on the voice . . . not subordinate to it," is "linear writing's contrary." From here, they take on inscription, representation, signification, and machinic questions of language and literature. "Alphabetical writing is not for illiterates, but by illiterates," they joke. "It goes by way of illiterates, those unconscious workers." And herein postliterate machineries are given articulation: "Writing has never been capitalism's thing. Capitalism is profoundly illiterate. The death of writing is like the death of God or the death of the father: the thing was settled a long time ago, although the news of the event is slow to reach us, and there survives in us the memory of extinct signs with which we still write . . . Once this is said, what exactly is meant when someone announces the collapse of the 'Gutenberg galaxy'?"[4]

Obviously, machineries are not exclusively "go to" concepts signifying material and immaterial phenomena or noumena. Machineries *are* phenomena and noumena, providing archetypes and prototypes that suggest how to act and think. Machinery refers to functional processes such as leveraging and to those more complex such as transporting, relaying, and oscillating or most simply to the quality or state of being machinic. Machinery is the facility to process and diffuse, which implies material, metaphoric, and metaphysical substrata.

LITERACIES

Albeit an ancient concept, literacy was nonetheless a nineteenth-century achievement born of illiteracy. "Illiteracy" antedated "literacy" by at least two centuries, with the latter coming into use as a derivative in the 1870s. "The quality or state of being literate," as literacy is first defined in the *OED* in 1908, through the end of the nineteenth century meant that one was "learned," "lettered," or "instructed in learning." Being literate (*litteratus*) was a pretext and meant one was prepared for learning. In Canada and the United States at the closing decades of the nineteenth century, census enumerators and immigration officials documented illiteracy rates by asking respondents "can you read?" and "can you write?" Literacy often merely meant an ability to write one's name or answering "yes" to these questions. Through the turn of the century, controversial laws in states such as Massachusetts required a potential voter to demonstrate competency by reading aloud short constitutional passages (50–100 words) to show that "he is neither prompted nor reciting from memory" and to copy a portion of the passage (10 words). For the most part, literacy came to mean that one could "read aloud intelligibly," eventually qualified as "read intelligently" and "write legibly." For instance, by the early 1920s, educators and psychometricians such as William McCall and Edward Thorndike had developed for New York state officials "functional silent reading" and "functional writing" tests to measure a respondent's "ability to express clear enough to be understood" "answers to certain questions upon the material being read" or copied (versus "oral reading of rare legal phraseology" and an "ability to copy a few words"). Legal challenges were brought against the new test, suggesting that it discriminated by measuring intelligence instead of literacy or innate instead of learned abilities. Charges included that literacy meant "a higher degree of education," but the high court ruled that it meant functional reading and writing abilities. Through the early twentieth century, like intelligence, literacy otherwise absorbed a much wider range of abilities, facilities, and faculties.[5]

Literacy was coextensive with behaviorism's provisional repudiation of phrenology, faculty psychology, and introspection, or, in effect, "the mind." Beliefs that "the mind is regarded as a machine of which the different faculties are parts," Thorndike reported in 1923, "have now disappeared from expert writings on psychology." He often referred to "the mind" in quotes, rejecting "magical effects," functions, and structures that educators and philosophers attributed to the organ. By the late 1920s, the following conclusion was common: "The mind is not a bundle of faculties, and also not a bundle of functions or, indeed, a *bundle* of anything—but rather a unitary psychic organism." More than anyone, including Bacon and Locke, Immanuel Kant established the mind as an aggregate of faculties, fundamentally reason (*vernunft*), understanding (*verstand*), and judgment (*urteilskraft*). He also joined these Aristotelian senses of animal, intellectual, and moral faculties with that of a faculty (*facultas*) as an art or branch of learning. In *The Conflict of the Faculties*, these senses are used nearly interchangeably. Somewhat ironically, Kant's elaboration underwrote materialist philosophy of mind from the 1780s, but trends turned toward physiological psychology. Around 1800, Franz J. Gall established craniology, renamed phrenology by his student Gaspar Spurzheim, as a science of mind identifying external cranial traces of internal cerebral faculties. Misguided by drawing discrete, practical terminations on the skull and brain, by the mid-1830s phrenology had mapped the mind into 35 distinct affective (propensities and sentiments) and intellectual (perceptive and reflective) faculties that could be cultivated. Like the soul chariot, the latter could be trained to control the prior; higher faculties could control the lower and therein separate the civilized from uncivilized, cultured from uncultured (humanity from animality, child from adult, etc.). Some protopsychologists at the time, such as Johann Herbart in 1816, rejected suggestions that faculties were innate and described them as a "hypothetical assumption": attempts to describe their "mutual influence in all their combinations" were "useless." Immensely popular in Britain and the United States, educators nevertheless gravitated to most features of the new psychology, which in the early 1840s expanded to 78 faculties, the basis of which developed differentially from child to adult minds.[6]

The constructiveness faculty was termed the *mechanical faculty* in 1648. In "Archimedes or Mechanical Powers," John Wilkins describes six mechanical faculties (balance, lever, wheel, pulley, wedge, screw) as if they are material, metaphoric, and metaphysical, like primordial, simple machines. "Books are not the only essentials," Francis Bacon wrote in *The Advancement*; machines and "instruments are required." Drawing from this tradition, phrenologists defined the constructiveness faculty in 1848 as "the making instinct and talent:" "manual dexterity in using tools; ingenuity; sleight of hand in constructing things, and turning off work, or whatever is done with the hands; disposition and ability to tinker, mend, fix up, make, build, manufacture, employ machinery and the like . . . Every human being uses it." Constructiveness was classified as a propensity, along with amativeness, combativeness, destructiveness, and secretiveness, actions that could be cultivated, indulged, or suppressed. As Herbart acknowledged, without control or restraint, constructiveness could manifest as an "organic excitation" or "disturbing element." Necessarily so, Maria Edgeworth advised in *Practical Education* in

1855, "parents are anxious that children should be conversant with Mechanics and with what are called the Mechanic Powers." Anxieties about constructiveness persisted, with schools of manual training, industrial art, and industrial education and institutes of technology first established and endowed in the United States during the latter half of the nineteenth century.[7] Manual training was defined with the exercise of faculties directly in mind as "training in thought-expression by other means than gesture and verbal language, in such a carefully graded course of study as shall also provide adequate training for the judgment and the executive faculty." Specialists drew from philosophers sympathetic to constructiveness, such as Pestalozzi and Froebel, as Calvin Woodward, founder of manual training in the United States, put it in 1884: "Watch the magic influence of a diet of *things* prescribed by the former in the place of words, and a little various practice in *doing*, in the place of *talking*, under the direction of the latter." Likewise, G. Stanley Hall, no fan of a discredited faculty psychology, reflected on the moral of the findings of his empirical child study: "Hence the need of objects and the danger of books and word cram." "Technological literacy" has roots in the constructiveness faculty, "mechanical intelligence," and in what the Massachusetts Commission on Industrial and Technical Education popularized in 1906 as "industrial intelligence," meaning "mental power to see beyond the task which occupies the hands for the moment to the operations which have proceeded and to those which will follow it—power to take in the whole process, knowledge of materials, ideas of cost, ideas of organization, business sense, and a conscience which recognizes obligations." Given modern machineries, schools were "too exclusively literary in their spirit, scope and methods."[8]

In one way, by the 1930s, the "old psychology" was discredited, but in another, the "new psychology" took up abilities, aptitudes, competencies, functions, intelligences, and literacies as nearly interchangeable with faculties. Certainly, functional psychology was accused of paraphrasing faculty psychology. The new dynamic psychiatry reinscribed drives, energies, forces, instincts, and mechanisms with alternate meanings. For example, mid-nineteenth-century assessments of constructiveness or the mechanical faculty became tests of mechanical ability and interest, or mechanical intelligence. The deployment of "industrial intelligence" was astute and part and parcel of an increase of intelligences in face of the waning currency of faculties. Educators and psychologists debated the existence and fragmentation of various intelligences versus a single, unitary "general intelligence." Thorndike had popularized multiple forms of intelligence in a 1920 *Harper's Monthly* article: there are "three intelligences," he confirmed, "which we may call mechanical intelligence, social intelligence, and abstract intelligence." Literacy, or specifically handwriting, had been reduced to dexterity, or "motor impulses," and an index of mechanical intelligence and machineries. Questions of how, what, or why conditions, functions, proclivities, and processes were "naturally innate" persisted. Hence intelligences had long fragmented into multiple types by the time Michael Youngblood conceived of "multiple intelligence" in 1979, which was subsequently exploited in Howard Gardner's *Frames of*

Mind in 1983. Literacies increased, beginning with "functional literacy" and then "economic literacy," coined in midst of the Great Depression. Ostensibly to counter the functional and respond to the great critique of the 1930s, "critical literacy" followed and was defined in 1943. "Scientific literacy" and "technical literacy" were coined in 1954; "media literacy" and "technological literacy" soon followed in 1961 and 1962. Technological literacy was productively fleshed out in the early 1970s through the Engineering Concepts Curriculum Project (ECCP) and an accompanying high school textbook titled *The Man-Made World*.[9]

Along with the singular "literacy," analysts began to routinely use the plural "literacies" by the mid-1970s. It was common to speak, in addition to those mentioned, of cultural, ecological, environmental, financial, social, and visual literacies. Literacies had long multiplied by the time "multiple literacies" was coined by Patrick Hartwell in 1985 and subsequently exploited by the New London Group in 1996. Nowadays, cyberliteracies, digital literacies, e-literacies, Information and Communication Technology and technoliteracies smoothly roll off the tongue along with about eighty other literacies.[10] Constructiveness is redistributed among a range of these old and new literacies. One does not need a special subliteracy to realize that each addition of a "new literacy," wherein the "new" has referred to a technological literacy since 1978, adds to a precarious edifice that in the conglomerate amounts to a new symbolic head.[11] The language faculty meant an ability to express our ideas verbally, and to use words that will best express our meaning; memory of words. It is now quite convenient to proclaim that the "virtual world is text" and go about our way, with business as usual so to speak, in cyberspace and game space. Indeed, game literacies, gaming literacies, and video-game literacies seemingly offer a new frontier, albeit back to the future, of literacies. Some archetypical literacies as established in literature include aboriginal, aesthetic, adult, business, cultural, electronic, financial, moral, pedagogical, scientific, theological and workplace, to name but a few.

Linguists and theorists rationalized the saturation and discovered a range of metaphors that hinted at reasons behind new literacies. For instance, in 1984, Sylvia Scribner tried to explain the trend: "Maximal literacy may begin for some through the feeder routes of a wide variety of specific literacies." At the base of "motivations for literacy," for Scribner, were three key metaphors ("literacy as adaptation, literacy as power, and literacy as a state of grace"). Whether metaphoric, or material, or both, others rediscovered that literacy was a commodity, medium, place, tool, technology, and a way of being. For the most part, there was discomfort in finding commodification and literacy as technology.[12] More than anyone, Walter Ong historically documented the materiality and metaphoricity of relations between the two, compressing common reactions into logic:

> Many would have it that the technologizing of culture also poses problems for professionalism. Technology implies machines. Machines are inhuman. Professionalism is human. And never the twain shall meet. Most such thinking is based

on a specious paradigm: a machine is taken to be an imitation organism—an animal that didn't quite make it. It thereupon triggers fear and resentment, for the machine obviously lacks the qualities that an organism should have: life, adaptability, moods and responsiveness to moods (which even many infrahuman animals clearly exhibit), adaptability to unpredicted change, and so on.

Resolving that literature is a machine, and acknowledging its machinic properties, was easier than accepting that literacy is a machine (e.g., for producing literacies). For example, a major theorist of the new criticism exclaimed in 1924 that "a book is a machine to think with," while William Carlos Williams introduced *The Wedge* in 1944 by asserting that "a poem is a small (or large) machine made of words." In the 1950s, Jacques Lacan extended insights of psychoanalysts on machines to symbolic systems. "The most complicated machines," he found, "are made only of words . . . The symbolic world is the world of the machine." Putting materiality to what he dubbed "hypermedia" and "hypertext" in 1965, Ted Nelson later outlined a comprehensive plan for designing "literary machines." These types of resolutions helped Deleuze and Guattari draw the material, metaphoric, and metaphysical from the "machine of expression" that Kafka used. Literacies and machineries overlap *and* differ, just as literature and machines are indistinct and analogical. Language and literacies are not exclusive to humans, and machineries are not limited to machines.[13]

Although La Mettrie boldly concluded that "man is a machine," he did not specify what type of machine. Once machines are given residual or "incipient vitality" and become metabolic, and once cultural and machinic evolution are coextensive with biological evolution, the types of machines that humans become necessarily change over time. After cybernetics in the 1940s, *homo communicans* became *machina processus*; speech became signal processing; reading became character recognition, information storage, and retrieval; and writing became word processing. In "A Mathematical Theory of Communication," Claude Shannon described the increasing indifference of machineries to literacies: frequently, "messages have *meaning*; that is they refer to or are correlated according to some system with certain physical or conceptual entities. These semantic aspects of communication are irrelevant to the engineering problem." A few years later, in 1950, Norbert Wiener clarified the indifference, explaining that "society can only be understood through a study of the messages and the communication facilities which belong to it . . . messages between man and machines, between machines and man, and between machine and machine . . . To me, personally, the fact that the signal [or communication] in its intermediate stages has gone through a machine rather than through a person is irrelevant and does not in any case greatly change my relation to the signal." Cyberneticians had cut the literary critics' critiques off at the pass: machines were getting good at "making the right noises for particular combinations of letters," and it does not matter whether or not they "understand what those noises mean." Librarians quickly yielded by accepting that "the 'information' in information theory is not the kind of information we mean when we talk of an 'informed' person." Literacies came face to interface with machineries. Recognizing implications, Marshall McLuhan exclaimed in 1962 that we "live in an electric or post-literate time" but nonetheless went on to say that "nothing could be more subversive of the Marxian dialectic than the idea that linguistic media shape social development, as much as do the means of production." Working from Marx's premise that "the hand-mill gives you society with the feudal lord; the steam-mill, society with the industrial capitalist," McLuhan reasoned that "antithetic periods, the Gutenberg and the Marconi or electronic," give you different literacies. The Gutenberg galaxy gives you "phonetic literacy," while the Marconi galaxy gives you "the new literacy," he insisted for years on end. Just as one could be machinic regardless of the literacies, one could be literate regardless of the machineries (analogical, new, digital, etc.).[14]

The world wants symbol systems and words—to make things exist, manifest magic, signify things, complement machineries, or at least incite discourse—but through semantic expansion or slippage, "literacies" are wont to overwrite nearly every discursive practice of machineries. Derrida's famous declaration that "there is nothing outside the text" offers a convenient medium through which literacies propagate and saturate and through which literate machineries seem subjugated and subdued. He nonetheless tried to moderate deconstruction's finding that the "world is text" by noting, "It was never our wish to extend the reassuring notion of the text to a whole extra-textual realm and to transform the world into a library by doing away with all boundaries." But this moderation was difficult or impossible given his defini-tion of *text* as "a differential network, a fabric of traces referring endlessly to something other than itself, to other differential traces." "Language contains its own inner principle of proliferation," cautions Michel Foucault.[15]

POSTLITERATE MACHINERIES

Apple CEO Steve Jobs's statement on literacy in early 2008 ("It doesn't matter how good or bad the product is, the fact is that people don't read anymore") provides a hint of the postliterate, as does a comment from a ventriloquized Bill Gates: "A generation or two has come along that can't be bothered to read; it absorbs all its information." These are hints to postliter-ate machineries. The postliterate does not merely refer to suggestions that literacy is no longer necessary or to a new era whereby illiterate children or savvy and selective kids can now cope without, given ranges of new media, prostheses and assistive technologies, as one commentator suggested (i.e., the new *postliterati*: "those who can read, but choose to meet their pri-mary information and recreational needs through audio, video, graphics and gaming"). Nor does it only refer to a time when the digital glow of crystal LEDs washes out the candlelight of the Gutenberg galaxy, a romantic turn of "secondary orality," electracy, or mediacy against what McLuhanites used to call "print-oriented bastards" (POBs). This is not quite the golden age of literacy yielding to the silicon age of postliteracies or a grand opening of the holographic or virtual *pharmakon*. It certainly does not simply refer to a de-alphabetization of the mind, as if it now could run on brainpower to codify and autocompile machine language. The postliterate points much less toward a waning state of literacy metaphors, practices, or skills and a cultural stage exceeding literacy than toward a recognition that machineries are no longer subordinate to literacies. The entire edifice is called into question with one modest insubordination—texting needs literacies; sexting needs something else.[16]

Naturally, it is tempting to answer "postliteracy" to the question "what comes after literacy?" This is sounder than saying that the newer and newest literacies come after the new literacies. Just as modernism has its literacies, postmodernism has its postliteracies. Postliteracy may be a key to decoding all the "posties," or postconditions, from postmodernism to transhumanism: "With fragmentation now the normal epistemological condition, and knowledge itself increasingly reduced to information which neither has to be memorised (as in oral culture) nor systematically catalogued (as in literate culture), with context downgraded, and audio-visual subsuming written communication, the most important 'post' of all comes into its own: post-literacy." It is attractive to describe an end of "technologically based dominance of texts over pictures" as "postliteracy," just as it is inviting to name this as the developmental stage following literacy. In this chapter, after asking what is a successor to "literacies," considering their saturation in a campaign of total literacies, it is also satisficing to answer "postliteracies." In postliteracies, does the day of rest occur after six days of work, is the moment of leisure after labor, the lifelong learning after schooling? In this, is the light, networked mind of dumb pipes after the heavy, overworked mind of intelligent letters? Even architects of "digital literacies," looking quaint and old-fashioned, have to be grandparented into the shadow of postliteracies.[17]

Inasmuch as the postliterate is commonly associated with the death or exhaustion of literature, it is from the exhaustion of literacies that postliterate machineries make sense and meaning. As Socrates suggests in the *Phaedrus*, preliterate machineries delivered *grammatikē* as somewhat dead on arrival—a "sacrifice to Mercury" as an ancient commentator phrased it. Bacon nonetheless felt that the "force, virtue, and consequences" of three inventions—namely, printing, gunpowder, and the compass—surpassed the wisdom of the ancients and "changed the face and condition of things all over the world." Analogical printing helped realize a proliferation of literate machineries. At the dawn of literacy at the nineteenth century's end, critics dreamed that a "literary machine could be invented for compressing fiction, like a cotton or hay press, by which the loose materials that make up the three-volume novel or the serial in twelve monthly parts could be packed into the dimensions of a pamphlet or of a good magazine story." The death or exhaustion of literature and the "literature of exhaustion," with its endless periods of anxieties and decline, is just preliterate, literate, and postliterate machineries at play and work. Unwittingly, literary theorist Harold Bloom summed up this dimension of postliterate machineries in a potshot at *Harry Potter*. "That's not reading because there's nothing there to be read." The best defense mustered thus far is dipping into the unconscious for dreams of twenty-first-century, on-demand, imachinic, and postmachinic literacies.[18]

Observing that no one reads anymore, Jobs could have stipulated that no one really reads or writes emails anymore. Yet, paradoxically, we have all become a postkid, postman, postwoman, or the gender-neutral and progressive postperson, "merely one stage in the long journey we might call the authorship of the self." Mobilities have turned us into email carriers anxious of being or going conative, postal, and postliterate, like Moses with the tablet, at any given moment. Fifty years ago and somewhat robotist, it seemed that cyborgs were destined to safeguard humans from machineries to leave them to their cherished literacies. Definitely not your grandmother's stationary literacies, cyborgs stamped a different postmark on the future and opted for text appeal *and* flash drive. God and goddess forbid that they would be slaves to literacies but not machineries.[19]

The postliterate resigns to a fact that digital pulses and analogical impulses are interdependent and coincidental—we are parts genuine and counterfeit, unadulterated and adulterated, lettered and machinic, the distribution of which is haphazard and designed. By and by, there has been a devaluation of currency in the "literacy myth" and "triumph of literacy" archetypes (i.e., civil, literate persons at the wit's end of uncivil, nonliterate machineries). It is uncertain whether the free will is there to do battle for literacies in the way one battled for literature. When the tide turns, literary technologies and technological literacies need redefining. Still, it will be a lot more meaningful to download previews of the future by spinning wheels of fortune and the world's horoscope, leveraging masses with mechanical muses, building bridges to unrealized destinations, riding chariots, motoring desires, or otherwise inciting and inventing non-, pre-, a-, and postliterate machineries. All things considered, I want to be preterliterate and machinic.

NOTES

1. Donna J. Haraway, "Manifesto for Cyborgs: Science, Technology and Socialist Feminism in the 1980s," *Socialist Review* 80 (June 1985): 65–108, on 101; Jay Lemke, "Metamedia Literacy: Transforming Meanings and Media," in *Handbook of Literacy and Technology*, eds. David Reinking, Michael C. McKenna, Linda D. Labbo, and Ronald D. Kieffer (Mahwah, NJ: Erlbaum), 283–301, on 283.

2. See, for example, Gilles Deleuze and Félix Guattari, *Anti-Oedipus: Capitalism and Schizophrenia*, trans. Robert Hurley, Mark Seem, and Helen R. Lane (New York: Viking, 1972/1977); Gilles Deleuze and Félix Guattari, *A Thousand Plateaus: Capitalism and Schizophrenia*, trans. Brian Massumi (Minneapolis: University of Minnesota Press, 1980/1987); Félix Guattari, *The Machinic Unconscious: Essays in Schizoanalysis*, trans. Taylor Adkins (Los Angeles: Semiotext(e), 1979/2011); Félix Guattari, *Chaosmosis: An Ethico-Aesthetic Paradigm*, trans. Paul Bains and Julian Pefanis (Bloomington: Indiana University Press, 1992/1995); John Johnston, *The Allure of Machinic Life* (Cambridge,

MA: MIT Press, 2008); Alistair Welchman, "Machinic Thinking," in *Deleuze and Philosophy*, ed. Keith A. Pearson (New York: Routledge, 1997), 211–29.

3. A much more elaborate history and philosophy of machineries can be found in Stephen Petrina, *Postliterate Machineries* (Unpublished Manuscript, 2012).

4. Deleuze and Guattari, *Anti-Oedipus*, 1–2, 9, 180, 183, 188, 208, 240; Deleuze and Guattari, "Psychoanalysis and Ethnology," *SubStance* 4 (1975): 170–97, on 185–86, 183, 190; "Balance Sheet—Program for Desiring Machines," *Semiotext(e)* 2 (1977): 117–35; Deleuze and Guattari, *A Thousand*; Félix Guattari, "On Machines," trans. V. Constantinopoulos, *Journal of Philosophy and the Visual Arts* 6 (1995): 8–12; Félix Guattari, "Machinic Heterogenesis," in *Chaosmosis*, trans. Paul Bains and Julian Pefanis (Bloomington: Indiana University Press, 1992), 33–57.

5. William P. Dillingham, ed., *Statistical Review of Immigration, 1820–1910* (Washington, DC: Government Printing Office, 1911), 84; "Literacy," *Oxford English Dictionary* (1888/1908); "Literate," in *An American Dictionary of the English Language*, ed. Noah Webster (New York, 1828). *Litteratus* in Roman antiquity also meant "branded into slavery." For changes to the *New York State Regents Literacy Test* and the history of voting literacy tests, see Arthur W. Bromage, "Literacy and the Electorate," *American Political Science Review* 24 (November 1930): 946–62; F. C. Crawford, "New York State Literacy Test," *American Political Science Review* 19 (November 1925): 788–90, on 789; Cayce Morrison, "New York State Regents Literacy Test," *Journal of Educational Research* 12 (September 1925): 145–55; "[Teachers] College News," "Notes from the Field," *Teachers College Record* 26 (1925): 584; J. Edgar Dransfield and Arthur Gates, "A Technique for Teaching Silent Reading," *Teachers College Record* 26 (1925): 740–52; "Triumph of the Literacy Law in New York," *Educational Review* 31 (January 1924): 40; "A New Literacy Test for Voters," *School and Society* 19 (March 1, 1924): 233–38; John R. Voorhis, "An Educational Test for the Ballot," *Educational Review* 31 (January, 1924): 1–4. For the history of literacy, see, for example, Jack Goody and Ian Watt, "The Consequences of Literacy," *Comparative Studies in Society and History* 5 (April 1963): 304–45; Harvey J. Graff, *The Literacy Myth: Literacy and Social Structure in the Nineteenth Century City* (New York: Academic Press, 1979); Ivin Illich and Barry Sanders, *The Alphabetization of the Popular Mind* (San Francisco: North Point Press, 1988); Carl F. Kaestle, "The History of Literacy and the History of Readers," *Review of Research in Education* 12 (1985), 11–53; Carl F. Kaestle, Helen Damon-Moore, Lawrence C. Stedman, and Katherine Tinsley, *Literacy in the United States: Readers and Reading Since 1880* (New Haven, CT: Yale University Press, 1991); Christine Pawley, "Information Literacy: A Contradictory Coupling," *Library Quarterly* 73 (October 2003): 422–52; Rene Wellek, "The Attack on Literature," in *The Attack on Literature and Other Essays* (Chapel Hill: University of North Carolina Press, 1982); and Raymond Williams, "Literacy," in *Keywords* (London, UK: Fontana Press, 1976).

6. Edward L. Thorndike, *The Psychology of Learning* (New York: Teachers College Press, 1913/1923), 363, 423; Herbert Sanborn, "The Dogma of Non-Transference," *Peabody Journal of Education* 5 (September 1927): 67–80, on 78. On Kant and faculties of the mind, see, for example, his *Critique of Pure Reason* and *The Conflict of the Faculties*, trans. Mary J. Gregor (New York: Abaris, 1798/1979); Jennifer Radden, "Lumps and Bumps: Kantian Faculty

Psychology, Phrenology, and Twentieth-Century Psychiatric Classification," *Philosophy, Psychiatry and Psychology* 3 (March 1996): 1–14. For the early history of phrenology, see David Bakan, "The Influence of Phrenology on American Psychology," *Journal of the History of the Behavioral Sciences* 2 (1966): 200–220; George Combe, *Outlines of Phrenology* (London, UK: Longman, 1835); "Phrenology," *North American Review* 37 (July 1833): 59–83; and "Phrenology [78 Faculties] and Animal Magnetism," *American Phrenological Journal* 4 (September 1842): 275–76. For Herbart's critique, see Johann Friedrich Herbart, *A Text-Book in Psychology*, trans. Margaret Smith (New York: D. Appleton, 1816/1891), 36–96, on 92; and G. F. Stout, "The Herbartian Psychology," *Mind* 13 (July 1888): 321–38.

7. Francis Bacon, *The Advancement of Learning* (New York: Collier, 1605/1902), 89; John Wilkins, "Archimedes or Mechanical Powers," in *Mathematical and Philosophical Works of the Right Rev. John Wilkins, Volume II* (London, UK: C. Whittingham, 1648/1802), 98; "Constructiveness—Its Definition, Location, Adaptation, and Cultivation," *American Phrenological Journal* 10 (January 1848): 22–25, on 22, 23, 24; "Education and Training Phrenologically Considered: Mechanical Talent and Skill," *American Phrenological Journal* 36 (July 1862): 16–17; Herbart, *A Text-Book*, 46; Maria Edgeworth, Practical Education.

8. "An Authoritative Definition of Manual Training," *Science* 13 (January 4, 1889): 9–10, on 10; Calvin M. Woodward, "The Fruits of Manual Training," *Popular Science Monthly* 25 (July 1884); 347–57, on 350; G. Stanley Hall, "The Content of Children's Minds on Entering School," in *Aspects of Child Life and Education* (New York: D. Appleton, 1921), 23; Carroll D. Wright, Warren A. Reed, and John Golden, eds., *Report of the Commission on Industrial and Technical Education* (New York: Teachers College Press, 1906), 5; and Charles Richards, "The Report of the Massachusetts Commission on Industrial and Technical Education," *Charities and the Commons* 16 (1906): 334–39, on 334. As one analyst remarked, "The very term 'manual training' suggests the now discredited 'faculty psychology' with its will-o'-the-wisp of general discipline." Jesse D. Burks, "Manual Activities in the Elementary School," *Elementary School Teacher* 11 (February 1911): 323–28, on 324.

9. Ruth M. Hubbard, "A Measurement of Mechanical Interests," *Pedagogical Seminary and Journal of Genetic Psychology* 35 (1928): 229–54; Charles Spearman, "'General Intelligence', Objectively Determined and Measured," *American Journal of Psychology* 15 (April 1904): 201–92; Edward L. Thorndike, "Intelligence and Its Uses," *Harper's Monthly Magazine* 140 (January 1920): 227–35, on 228; Max Freyd, "The Personalities of the Socially and Mechanically Inclined," *Psychological Monographs* 33 (1924): 1–101, on 14; Michael S. Youngblood, "A Nonverbal Ability Test," *Studies in Art Education* 20 (1979): 52–63, on 52; Ralph B. Guinness, "Critical Literacy," *Social Education* 7 (April, 1943): 165–66; E. R. Purpus, "Scientific and Technical Literacy," *Journal of Higher Education* 25 (December 1954): 475–78; Thomas Liao and Emil J. Piel, "Toward Technological Literacy," *Engineering Concepts Curriculum Newsletter* 6 (1970): 2–4; Thomas Liao, Emil J. Piel, and John Truxal, "Technology-People-Environment: An Activities Approach," *School Science and Mathematics* 75 (1975): 99–108; Engineering Concepts Curriculum Project (ECCP), *The Man-Made World* (New York: McGraw-Hill, 1971). The ECCP was funded by the US National Academy of Engineering (NAE) and

National Science Foundation (NSF). *Technological literacy* was defined as an understanding of "the nature, the capabilities, the limitations and the trends of technology." *The Man-Made World*, despite its sexist title, was quite an amazing text, integrating the themes of technology, people, and the environment into a wide range of activities and lessons.

10. James Collins, "Literacy and Literacies," *Annual Review of Anthropology* 24 (1995): 75–93; George Steiner, "Text and Context," *Salmagundi* 31/32 (Fall 1975–Winter 1976): 173–84, on 179; Patrick Hartwell, "Grammar, Grammars, and the Teaching of Grammar," *College English* 47 (February 1985): 105–27, on 123; and The New London Group, "A Pedagogy of Multiliteracies: Designing Social Futures," *Harvard Educational Review* 66 (1996): 60–92. On "new literacy" and "new literacies," see, for example, Joshua Lederberg, "Digital Communications and the Conduct of Science: The New Literacy," *Proceedings of the IEEE* 66 (November 1978): 1314–19; and Brian Street, "New Literacies in Theory and Practice: What Are the Implications for Language Education?," *Linguistics and Education* 10 (1998): 1–24. For analyses of this saturation, see Vartan Gregorian, "Education and Our Divided Knowledge," *Proceedings of the American Philosophical Society* 137 (December 1993): 605–611; Vartan Gregorian, "Technology, Scholarship, and the Humanities: The Implications of Electronic Information," *Leonardo* 27 (1994): 129–33; and Anne F. Wysocki and Johndan Johnson-Eilola, "Blinded by the Letter: Why Are We Using Literacy as a Metaphor for Everything Else?," in *Passions, Pedagogies, and 21st Century Technologies*, eds. Gail Hawisher and Cynthia Selfe (Logan: Utah State University Press, 1994), 349–68, on 360. Counting 197 literacies, Wysocki and Johndon-Eilola ask, "Why aren't we instead working to come up with other terms and understandings—other more complex expressions—of our relationship with and within technologies?"

11. For definitions of the language faculty, see "Definition of the Faculties According to Their Numbers," *American Phrenological Journal* 20 (July 1854): 24. I am included among those helping to proliferate and saturate new literacies. See, for example, Stephen Petrina, "The Politics of Technological Literacy," *International Journal of Technology and Design Education* 10 (2000): 181–206; Stephen Petrina, "Review of *The Civilization of Illiteracy*," *Journal of Technology Education* 11 (2000): 69–70; Stephen Petrina, "Human Rights and Politically Incorrect Thinking versus *Technically Speaking*," *Journal of Technology Education* 14 (2003): 70–74; Stephen Petrina, *Advanced Teaching Methods for the Technology Classroom* (Hershey, PA: Information Science, Inc., 2007), 223–50; Stephen Petrina and Ruth Guo, "Developing a Large-Scale Assessment of Technological Literacy," in *Assessment in Technology Education*, eds. Marie Hoepfl and Michael Lindstrom (New York: Glencoe-McGraw Hill, 2007), 157–80; and Stephen Petrina, Oksana Bartosh, Ruth Guo, and Linda Stanley-Wilson, "ICT Literacies and Policies in Teacher Education," in *The Emperor's New Computer*, ed. Tony Di Petta (Rotterdam: Sense, 2008), 89–109. But also see the analysis of machineries in Karen Brennan, Franc Feng, Lauren Hall, and Stephen Petrina, "On the Complexity of Technology and the Technology of Complexity," in *Proceedings of the Fourth Complexity Science and Educational Research Conference*, ed. Brent Davis (2007), 47–73.

12. Sylvia Scribner, "Literacy in Three Metaphors," *American Journal of Education* 93 (November 1984): 6–21, on 18, 8; Terry Beers, "Commentary: Schema-Theoretic Models of Reading: Humanizing the Machine," *Reading Research*

Quarterly 22 (Summer 1987): 369–77; Colin Lankshear and Peter O'Connor, "Response to 'Adult Literacy: The Next Generation,'" *Educational Researcher* 28 (January–February 1999): 30–36; Elizabeth Birr Moje, Allan Luke, Bronwyn Davies, and Brian Street, "Literacy and Identity: Examining the Metaphors in History and Contemporary Research," *Reading Research Quarterly* 44 (October 2009): 415–37; Constance A. Steinkuehler, Rebecca W. Black, and Katherine A. Clinton, "Researching Literacy as Tool, Place, and Way of Being," *Reading Research Quarterly* 40 (January–March 2005): 95–100; and John M. Willinsky, "The Seldom-Spoken Roots of the Curriculum: Romanticism and the New Literacy," *Curriculum Inquiry* 17 (Autumn 1987): 267–91.

13. Walter J. Ong, "Presidential Address 1978: The Human Nature of Professionalism," *PMLA* 94 (May 1979): 385–94, on 392; Walter J. Ong, *Orality and Literacy: The Technologizing of the Word* (New York: Routledge, 1982/1988); Walter J. Ong, "Reading, Technology, and the Nature of Man: An Interpretation," *Yearbook of English Studies* 10 (1980): 132–49; I. A. Richards, *Principles of Literary Criticism* (New York: Harcourt), 1; William Carlos Williams, "Author's Introduction," *Collected Poems of William Carlos Williams, Volume 2 1939–1962*, ed. Christopher MacGowan (New York: New Directions, 1944/1964), 54; Theodor Holm Nelson, *Literary Machines* (San Antonio, TX: Author, 1981/1987); Gilles Deleuze and Félix Guattari, "What Is a Minor Literature?," trans. Robert Brinkley, *Mississippi Review* 11 (Winter/Spring, 1975/1983): 13–33, on 18; and Gilles Deleuze and Félix Guattari, *Kafka: Toward a Minor Literature*, trans. Dana Polan (Minneapolis: University of Minnesota Press, 1975/1986), 18–19. One machinic literature, see also, for example, Michael Heim, *Electric Language: A Philosophical Study of Word Processing* (New Haven, CT: Yale University Press, 1987); Jeffrey Masten, Peter Stellybrass, and Nancy Vickers, eds., *Language Machines* (New York: Routledge, 1997); Brian McHale, "Poetry as Prosthesis," *Poetics Today* 21 (Spring 2000): 1–32; and Cicelia Tichi, *Shifting Gears: Technology, Literature, Culture in Modernist America* (Chapel Hill: University of North Carolina Press, 1987).

14. Julien Offray De La Mettrie, *Machine Man and Other Writings*, trans. Ann Thomson (Cambridge: Cambridge University Press), 39; F. F. Blackman, "Incipient Vitality," *New Phytologist* 5 (January 1906): 22–34; Claude E. Shannon and Warren Weaver, *The Mathematical Theory of Communication* (Urbana: University of Illinois Press, 1949), 31; Norbert Wiener, *The Human Use of Human Beings: Cybernetics and Society* (London, UK: Free Association Books, 1950/1989), 16; Peter Sanders, "Reading and the English Teacher: Toads Being Good," *English Journal* 63 (September 1974): 59–60, on 59; Mortimer Taube, "Documentation, Information Retrieval, and Other New Techniques," *Library Quarterly* 31 (January 1961): 90–103, on 92; Lester Asheim, "Introduction: New Problems in Plotting the Future of the Book," *Library Quarterly* 25 (October 1955): 281–92; Marshall McLuhan, *The Gutenberg Galaxy: The Making of Typographic Man* (Toronto: University of Toronto Press, 1962), 2, 144; Marshall McLuhan, *Understanding Media: The Extensions of Man* (New York: McGraw Hill, 1964), 49, 316; and Karl Marx, *The Poverty of Philosophy* (London, UK: Martin Lawrence, 1847/1955), 92, 112.

15. Jacques Derrida, *Of Grammatology*, trans. Gayatri C. Spivak (Baltimore: Johns Hopkins University Press, 1976), 158; Jacques Derrida, "Living On: Border Lines," trans. J. Hulbert, in *Deconstruction and Criticism*, ed. Harold Bloom (London, UK: Routledge, 1979), 75–176, on 84; and Michel Foucault, *The*

Order of Things: An Archaeology of the Human Sciences (New York: Vintage, 1970/1994), 40. Derrida later clarified "that one never accedes to a text without some relation to its contextual opening and that a context is not made up only of what is so trivially called a text." Jacques Derrida, "Biodegradables: Seven Diary Fragments," trans. Peggy Kamuf, in *Critical Inquiry* 15 (1989): 812–73, on 841. To be sure, postmodernism came a bit late, as Saint John had some time ago edited *Genesis* to shift from *Old* to *New Testament*: "In the beginning God created the Heaven and Earth . . ." became "In the beginning was the Word." Although the word now preceded the flesh in the world, Logos and logos, World and word, still seem coemergent if not interchangeable.

16. Jobs is quoted in J. Markoff, "The Passion of Steve Jobs," *New York Times* (January 15, 2008); Gates is ventriloquized by Updike, quoted in Kristóf Nyíri, "The Humanities in the Age of Post-Literacy," *Budapest Review of Books* 6 (1996): 110–16, on 110; D. Johnson, "Libraries for a Post-Literate Society" (2008): Retrieved February 3, 2010, from http://doug-johnson.squarespace.com/blue-skunk-blog/2008/8/13. For "secondary orality," see Ong, *Orality*, 136; Illich and Sanders, *The Alphabetization*, 106–27; and Barry Sanders, *A Is for Ox* (New York: Pantheon, 1994). For the POB quote, see John Barth, "The Literature of Exhaustion," in *The Friday Book: Essays and Other Non-Fiction* (London, UK: Johns Hopkins University Press, 1984), 62–76, on 71.

17. Dave H. Ravindra, Adama Ouane, and Peter Sutton, "Editorial Introduction," *International Review of Education* 35 (1989): 383–87, on 383; Dave H. Ravindra, Adama Ouane, and Peter Sutton, "Issues in Post-Literacy," *International Review of Education* 35 (1989): 389–408; Sean Regan, "Postmodern Tenor," *AQ: Australian Quarterly* 71 (September–October 1999): 6–9, on 8; Nyíri, "The Humanities," 113; Kristóf Nyíri, "Post-Literacy as a Source of Twentieth-Century Philosophy," *Synthese* 130 (February 2002): 185–99; and Kristóf Nyíri, "The Networked Mind," *Studies in East European Thought* 60 (June 2008): 149–58.

18. Lucian, "The Illiterate Bibliomaniac," in *Lucian of Samosata, Volume II*, trans. William Tooke (London, UK: Longman, 182), 513–30, on 513; Bacon, *Novum Organum*, 300; "Review of *Good Bye, Sweetheart*," *North American Review* 115 (October 1872): 435–37, on 435. For the death of literature, see Barth, "Literature of Exhaustion"; J. Yellowlees Douglas, *The End of Books—or Books without End? Reading Interactive Narratives* (Ann Arbor: University of Michigan Press, 2000), 1–10; Kathleen Fitzpatrick, "The Exhaustion of Literature: Novels, Computers, and the Threat of Obsolescence," *Contemporary Literature* 43 (Autumn 2002): 518–59; Welleck, "The Attack"; and Harold Bloom, "Interview with Harold Bloom," *The Charlie Rose Show* (Program #2723), PBS WNET (New York, July 11, 2000). On the postmachinic, see Ronnie Lippens, "Imachinations of Peace: Scientifictions of Peace in Iain M. Banks's *The Player of Games*," *Utopian Studies* 13 (2002): 135–47, on 146; and Laurence A. Rickels, "Half-Life," *Discourse* 31 (Spring 2009), 106–23, on 107, 108.

19. On the postperson, see Kedrick James, "Writing Post-Person: Literacy, Poetics, and Sustainability in the Age of Disposable Discourse" (PhD diss., University of British Columbia, 2008), 3; and Allen Buchanan, "Moral Status and Human Enhancement," *Philosophy and Public Affairs* 37 (2009): 346–81. For the cyborg future, see Manfred E. Clynes and Nathan S. Kline, "Cyborgs and Space," *Astronautics* (September 1960): 26–27, 74–76, on 27.

The Hidden Voice of Youth

CHAPTER 3

TECHNOLOGY AND
TECHNOLOGY EDUCATION

PERSPECTIVES FROM A YOUNG PERSON

Molly Watson

Editor's Note:

Every author who has contributed to this book believes that education in general, and technology education in particular, should not in any way be about training young people. Rather, the authors believe that education should be about the development of critical faculties—critical faculties that enable young people, and people in general, to challenge the received wisdom of the day. However, with very few exceptions, technology education around the world is perceived to be an area where young people are trained to understand an already-established technological world where all knowledge is stable and fixed. This type of training tends to manifest itself in a curriculum that favors the development of preestablished skills. These particular skills bear very little resemblance to the technologically mediated world that we all occupy today and have little to do with technological literacy. Several chapters in this book, in their own way, challenge the extant curriculum as being outdated, outmoded, and having nothing to do with the technological world we inhabit. Molly Watson is a 15-year-old school student. Her chapter, I would argue, resonates with many of the perspectives offered in the other chapters. Molly's chapter details her perception of technology and what she perceives school-based technology education offers her as a student. There appears to be no match whatsoever.

Her chapter is not academic in the conventional sense. However, it offers a personal account of what appears to be a mismatch between technology education today and the technological world that young people occupy. It is both poignant and illustrative of the issues raised in the other chapters. It grounds the academic rigor offered elsewhere in this book with a realistic account by a typical young user of technology. It does not offer a universal perspective; no account ever can. As editor of this collection, I believe that this chapter adds a unique dimension to the book.

Every generation has experienced some form of technological development that affects them to some degree. I am of a generation that that was born into a new age of digital communication. As a 15-year-old, I don't remember when computers weren't around or when mobile phones weren't commonplace. Such things were revolutionary only a couple of decades ago but have quickly become something that's simply part of the everyday, and there's an assumption that everyone should own these gadgets, which are more like necessities now. Technology is not something that's really questioned anymore; it's just something that is *there*, a part of our environment and our society. It surrounds us 24/7, and we thrive on it individually and collectively—but we don't often realize just how much we use it and, in some cases, depend on it. It's almost subliminal. We tend to treat technology as just an everyday feature without really knowing the first thing about the power it can have, or certainly its colossal impact on the world. It is an impact that, I feel, affects people my age more than anyone.

Technology is absolutely rife among my peers—both the use and abuse of it. In many cases, young people are far more confident than adults when it comes to working gadgets, and this is something that can be seen all around us in households and schools. I remember when our first family desktop computer was installed in our home. My brother was at elementary/primary school, and I was even younger. Yet we both were confident around it for a long time before our parents dared log on. Similarly, this can be seen in the classroom, where technology is being incorporated more and more into our lessons. Although I do think this is a good idea, since technology is becoming a huge part of our society, sometimes it doesn't work out. The students actually end up teaching the teachers, who sometimes tend to be far more uncertain of technology than we are. I find this to be quite a strange thing because, on the whole, adults are considered far more knowledgeable and capable than children. Yet, surprisingly, when faced with a machine too huge and complex to ever really explain or control, filled to the brim with endless information, it's the children that often seem to handle it the best while the adults may struggle to grasp the basics. I think it all has to do with adapting. If a child is brought up surrounded by something, then he or she is more likely to feel confident and familiar with it. Infants are almost being fed as much technology as they are milk, so it's hardly surprising that they won't feel uncertain around computers and phones.

After all, nowadays technology fills in so many areas of our everyday lives. We're exposed to almost everything because of technology, and it's so much easier to access whatever we want. One of its major advantages is enabling users to download music, films, and books onto iPods and Kindles. There's no longer any need to go to the shops to buy a song, and downloading music is much cheaper and more space-efficient. I wouldn't be listening to half the music I listen to if I'd had to go out to buy it in a shop—it's considered the norm to have hundreds, or even thousands, of songs stored on an iTunes account, which would take up a whole lot of room if they were physical CDs! There's no denying that downloading is very practical and efficient. However, strangely enough, the ease of downloading can sometimes have the opposite effect: people I know have returned to buying record players because they *prefer* the rusty, flawed sound of vinyl and the challenge of building up a record collection to downloading music. Equally, I prefer holding a book in my arms and turning the pages by hand to reading from a screen. There's something very clinical about Kindles and iPods; I think it's fair to say they lack character. But it's all about preference; they definitely have their place and have turned a generation on to literature and music. When it comes to Kindles, I think I'm an exception to the rule in my preference for books: the vast majority of my peers would take the sleekness and practicality of a Kindle, despite its clinical feeling. I find it's often the same with teenagers and children; on the whole, we *like* technology more than adults. We tend to opt for the more high-tech option, and I think that's because we were born comfortable with it.

Of course, for today's adults, it's a lot harder to adapt as well as children do. Advanced technology is still a relatively new concept; it's a change, and change can make people feel uneasy. So, approached from different angles, technology can mean different things—while adults may see it as a daunting new concept beyond their control, young people may see it as entirely the opposite. In fact, young people (myself included) don't really see it as "technology" at all!

I think the very word "technology" has always been something associated with newness, learning, and development—something that requires skill and concentration, almost a challenge. But among people my age, this isn't what I see. I see all gadgets—be they mobile phones, iPods, or Kindles—being used as though they are the most ordinary things in the world. I realize that to some people these things are the height of high-tech, but the truth is, in the teenage world, there's a whole different perspective. Technology isn't used just as technology; it's also used as a status symbol. When I had just started high school, I remember the boys and girls in my class having a conversation about who had what kind of phone. They were all listing the best brand names—BlackBerrys, iPhones, and other variations—except there was one girl who didn't have a "good enough" phone. She felt the need to lie about it and just pretended because that was easier than admitting her mobile wasn't like everyone else's. Later on, when her friends saw her real phone, she threw it on the ground so she could tell her mother it was broken and she needed a new one. This competitiveness isn't out of the ordinary in my high school, and I'm sure it isn't in any other. Technology is a way of saying who you are, and everyone wants his or her peers to get the right idea. It's like a competition, sometimes, on a par with who has designer clothes and who wears the most cosmetics.

This is when technology starts to show its ugly side as something that can be abused—when it opens up a gateway to a whole new level of bullying, known as cyberbullying. Many people can underestimate the effect of cyberbullying, and other people are quite oblivious to it. Cyberbullying is incredibly common with people my age, and I see a huge amount of it from day to day. Social networking sites such as Facebook are very popular, especially among teens; the vast majority of people in my year at school have an account. These sites do undoubtedly have their good sides. As a member of Facebook myself, I know that it proves to be a great way of keeping in touch with friends and finding information about the people and things that interest you. However, it can definitely have its downsides as well. Bullying has always been around in schools, so of course that in itself is nothing new. What's made bullying these days scarier than ever is the ability to make it far more public and the inability to escape from it. Gone are the days when a victim was bullied only within the school gates. Today, the bullies can catch up with you from the comfort of their own homes while hiding behind a screen for protection. Certain websites even provide the optimum opportunity to make people feel small. Current fads like "ask.fm" and "spillit.me" allow you to create a webpage where *anybody* can anonymously ask you questions or say what they really think of you. This practically encourages criticism and victimization of others, and sadly these are both things that happen a lot among young people. Even people who appear to be very popular at school receive some disgusting messages and threats through these websites because it's just so easy. All it takes is somebody to type out a few vicious words under the comfort of complete anonymity. In the past few months, there have been a couple of cases in which young teenagers have taken their lives due to the threats and harassment they have been given on "ask.fm."

On the other hand, the ease of reaching people over the Internet can be a good thing. Speaking from experience, I know getting in touch with people over the Internet can provide support and advice without the pressures of face-to-face speech. I've suffered from illness in the past, which can be very isolating, and joining a forum designed for other young people in my situation definitely helped me feel less alone, as I didn't know anyone in person who was going through the same thing I was. Technology gives people this chance, which was never available before, to be able to communicate with anyone at all. It connects people from all around the world, and no one has to be entirely alone, which is a great thing. Obviously, when you've never seen people face-to-face, you have to be careful not to give out personal information or to become attached to or obsessed with them, as they may be completely different in reality from who they are on a computer screen. But sharing experiences, asking advice, or just chatting is harmless enough and can provide a safe method of socializing for those who may not have the opportunity or ability to do so otherwise.

In addition to connecting with people individually, nowadays it is easier than ever to reach wider audiences through technology. Websites and forums can provide excellent platforms for people who want their voices to be heard or to share a talent. Countless singers who are now world famous started out by uploading videos of their songs to YouTube, where so many people would watch them that word spread to talent scouts. The same can be said for most other hobbies and interests. Whether you're into writing, acting, photography, or athletics, there's bound to be a website on which you can launch your career and get noticed. All this means that our dreams and aspirations are more within reach than ever before, and that can only be a good thing.

In fact, it could be argued that it's almost *too* easy to become an Internet sensation—even for people who completely lack talent. The truth is, a number of people in this day and age are fame hungry and will resort to almost anything for a moment in the spotlight. One hundred hours of videos are uploaded to YouTube every *minute*—that's more than four days of material uploaded for the world to see every sixty seconds. Huge proportions of these people are contributing nothing of value but are instead taking the risk of being publicly humiliated just to get their 15minutes of fame. In some cases, they do become the "flavor of the month," but often for all the wrong reasons: people laughing at them and, more often than not, giving them a considerable amount of hate comments. I often use YouTube, as I find it to be an excellent method of listening to music or watching videos that interest me. In fact, I've even uploaded a video project I did with friends. Making the video was something that we all enjoyed, and meeting up to film and plan it together brought us closer together. It was a great feeling to see the end result of a lot of hard work. We only did it for fun, but upon uploading it to YouTube, we got a few thousand hits in a short amount of time. On the whole, the comments we received were positive, but inevitably we also got a few malicious ones. It didn't bother me all that much, mainly because I expected it. Sadly, it's pretty much a given that you'll get some negativity on YouTube, no matter what you upload. But some people can be more sensitive than this and find it hard to deal with the hate that can go hand-in-hand with Internet fame. I've found that if you're going to put yourself out there on the Internet, you have to be thick skinned by nature.

I think there's definitely a lot of naivety surrounding such technological powers as the Internet. I know several people who have thrown house parties, and the amount of guests got out of hand because these events were leaked on Facebook or some other social networking site. Sometimes one of the guests openly advertised it, and other times it was the hosts themselves that made a mistake and posted about it publicly. The thing is, often people just don't seem to realize how potent technology can be. Before you know it, something that was supposed to be private can spread until everybody knows about it. But people continue to risk sharing what really should stay completely private. There's always one thing or another circulating. Just among my own peers, mobile numbers have been revealed, embarrassing videos have been uploaded, and even topless photos have been passed around; some people don't seem to learn from it. A sobering truth is that once something is posted online, no matter what it is, it can never be removed completely. Even once it's "deleted," it will still be floating around somewhere in cyberspace, and this is something that can often come back to bite people in the future when they apply for jobs and universities. This is a fact that very few people know, but I think it should be the first thing we're taught when talking about technology—people don't fully grasp that technology is in fact beyond our control. I find this a very strange thing, because all technology is manmade, dating back to ancient history. Technology has always been around in some capacity—even the wheel and the axe were once examples of technological equipment—something that we made so we could use it and manipulate it to our convenience. The difference between "old" and "new" technology is that over the years it has become more and more complex until it peaked with the birth of the Internet. This is technology so advanced that it knows no limits. It's something so incredibly powerful that it almost controls *us*, rather than the other way around, and we don't even know everything it does or what it means. Simply put, even though we made it, we don't understand it, and this is a frightening idea to me. If a person is unaware of what they are getting into when they post things on the internet, should they be posting things at all?

The rule that "what goes online stays online" really starts to hit home when you think about all those millions of people that have a Facebook account—myself, and the majority of people I know, included. To state the obvious, none of these millions are going to go on living forever. In the bluntest terms possible, we're all going to die. So this may well leave you wondering just what will happen to these millions of accounts when their possessors are no longer alive. The answer, in short, is that the accounts will stay there pretty much indefinitely, just the same as they were when the people who held them were alive. If you happen to have told somebody your password, then that person could deactivate your account, but that just means that it lies in wait like a dormant volcano: as soon as somebody enters your old email address and password, it's reactivated, just like that. If nobody else knows your password, then your account will simply sit there open as it always was. Depending on your outlook, this may seem like a terrifying prospect: having all your old photographs and everything you've written just floating around as if you'd never died. It can seem very eerie, almost "sci-fi." However, there are positives, too. In a sense, it immortalizes a piece of you: it's something people can look back on to remember you by, and it's a place where people can write messages and tributes to you if it helps them cope. I suppose it's kind of like an online memorial—a legacy of sorts. How you feel about the situation completely depends on how you see it.

I think that explains all these opposing views on technology. Quite simply, it depends on how you see it. It's a strange situation I'm in as I write this, because although I'm just a teenager, I think I've seen more technology than most of the adults I know. It's something that I am constantly using or witnessing the use of, even when I'm not fully aware of it, and it is something I think I've been conditioned to know pretty well. I understand that some people find technology frightening and new, and I know I'll probably be the same when I'm older and something comes along that I can't identify with. It's always been the same, really—technology isn't considered new technology for long. It's new technology for as long as it's unfamiliar and needs to be understood. I'm not claiming that *I* understand technology, because I don't think anybody truly does. What I do understand is how it affects me and the people around me. If you take a look back at every paragraph I've written in this chapter, you'll see that essentially they're about the people behind the technology: how people use it or misuse it, as the case may be; how it makes people feel; what it makes people say or do. These are the things I know, because I see them every day. Whatever the technology is, as basic as the wheel or as complicated the Internet, it's all about what people can do with it. I believe technology is what we make of it, the pros and cons of it—we're responsible for it all. I've found some aspects of technology that are very ugly and some aspects that are equally brilliant, and it's up to us which one of these we make it. Because, when you think about it, what is technology without the people who use it?

I do not think my perspective on technology is atypical for people my age. This is why I find it strange when I consider how we are educated about it. What we are taught about technology and the way in which we are taught about it seem like pieces of the puzzle that do not really fit. In my school, we have two main forms of what is considered "technological education." These are called CDT (Craft and Design Technology) and ICT (Internet and Computing Technology). They both have very different connotations surrounding them. CDT is a subject many people in my school see as an "easy ride" and "unimportant." It is not considered at all academic, even by many of the teachers, and nothing much is done to stop this. ICT, on the other hand, is considered academic. This confuses me somewhat, because I cannot help thinking that outside the school gates, computing is something nearly everybody does confidently and capably, while construction is a task about which we would be far less certain. So why are these two forms of technological education thought of so very differently? Because when I strip them both down to what they are basically teaching, they seem very similar to me. CDT is teaching you how to make things that you can use. ICT is teaching you how to use things that have been made. You use your hand in CDT. You use your mouse in ICT. Neither of them seems more academic than the other, when I put them in those essential terms. They are both just teaching "how to." Whether it is how to build a bookshelf or how to work specific software is irrelevant. Technology education, in either form, is currently about giving an instruction or formula with the student following it precisely.

I personally do not think either CDT or ICT is irrelevant. They both matter, depending on what career path one wishes to go down. What I disagree with, however, is how they are presented, because in my opinion, we are predominantly taught to remember things and not necessarily to comprehend, appreciate, or identify with them. If we are given a task to do on a computer program—for example, Photoshop—then we are told how to make one specific change to one specific picture and not how to actually use Photoshop.

We now know how to blur a line on a photo of a donkey, but that is where our knowledge ends; we still cannot use that program because we haven't been taught how to. In the next lesson, we will move onto another program, and so it will continue—getting only glimpses of all the things we could be learning. Technology education has become far too specific, in my opinion. I feel we are not encouraged to use our initiative but rather are told what to make and what to do. It seems to me that we are given little explanation for *why* we are learning what we are learning, *why* these programs exist, and *why* they could be useful to us in the future. We barely seem to be told what it is we are learning. Reasons do not seem to come into the learning process. We are given a sheet with a method, and we often do not know what we are learning about; we are just memorizing a formula. How is this effective education? I asked a friend of mine, Alice, how she felt about her education in the field of technology. This very important to Alice, who is one of very few girls in school who chooses to study CDT and ICT and who wants a career in the more hands-on aspect of technology. She felt very strongly that the curriculum being taught left much to be desired. "CDT is beginning to focus far too much on computers," she told me.

> I am passionate about the handcrafts, and I feel as though they are being completely lost. We have two types of technology in the school, and they are supposed to offer two different types of skills. But more and more in what is supposed to be "Craft and Design," we sit in front of a computer. The worst part is, we don't know why we are doing what we are doing, or sometimes even what we should be doing. It seems to me that I simply press buttons that I don't even know the purpose of. All I want to do is make things using my own brain and my own two hands, not an artificial brain and an artificial hand. But that doesn't seem to me to be what happens.

I believe technology education would be far *more* effective if it were open, because it currently seems to be very narrow in terms of what it offers a student. It is clear to see that from what Alice experiences in her CDT lessons. In the real world, technology is not about doing what you are told. It is about using things to *your* advantage and having your own ideas. Technology is actually a very free thing, but this does not seem to be the way it exists in schools. We do not seem to be developing a real understanding of technology, and this can only lead to a lack of confidence and initiative—things that schools claim to "build up" in young people. I don't think it would be difficult to greatly improve technology curricula for pupils, in both the CDT and ICT areas. But currently, the curriculum does not seem to lead us to be self-sufficient or to use our brains and creativity. I think that needs to change. There is too much focus in the current technology curriculum on following orders. Most young people I know do not want this. And perhaps the reason many teenagers dislike school is because they are not given sufficient opportunities to express themselves, and they are not presented with things they feel are relevant. And if something is relevant, they are not always told why it is relevant. I think in technology, it is *crucial* that we can express ourselves and that we are told "why." I have already stated that many people abuse technology, but perhaps if, along with using it correctly, we were encouraged to discuss the consequences of what can happen when we do not, this would not be an issue. How can we be expected to be responsible and knowledge-able when there is an insufficient focus on discussing the dangers? And how can we be expected to turn into successful, independent individuals when the main emphasis in learning appears to be on following commands?

ISSUES RELATING TO DEMOCRACY

CHAPTER 4

TECHNOLOGICAL LITERACY AND DIGITAL DEMOCRACY

A RELATIONSHIP GROUNDED IN TECHNOLOGY EDUCATION

P. John Williams

INTRODUCTION

The history of technology education has been characterized by continual reinvention; because the nature of technology is dynamic, the nature of technology education also should be dynamic. Current developments in digital communication technologies present further opportunities for technology education to contribute to dispositions in students that are fundamental to their participation in the developing global digital democracy. The nature of the global context, within which the exercise of democracy takes place, warrants discussion.

GLOBALIZATION

In its broadest sense, *globalization* refers to the effects of recent significant changes that have occurred in the international economy. These include the demise of communist states and the spread of capitalism; the increasing mobility of capital, labor, and goods and services; acceptance of market forces; new international divisions of labor and a diminished role for the state; and the international network of financial institutions in which decisions made in one sector of one country have a ripple effect of influence throughout the world.

These factors have resulted in a homogenization of production, consumption, and cultural values across the world (United Nations, 1995).

However, there are all kinds of qualifiers to this perception of globalization. A large (though diminishing) proportion of the world's population remains out of touch with globalization's influences. As communication networks become more pervasive, ubiquitous, and instrumental within the forces of globalization, the "balance of power" shifts to those who are able to participate in these networks, which may be counterbalanced by those who control the networks. So as Keirl (2003, p. 57) points out, globalization "can be seen as aggressive or benign, overt or covert, welcomed or loathed, and that one's perceptions are very much a matter of politics or place in the whole affair."

Some would posit that the outcome of this continuing trend of globalization, which so far has as a common theme the sublimation and domination of the third world, is an elimination of the boundaries between the traditional categories of first, second, and third world (Dirlik, 1997). Unfortunately, this does not mean that poverty, inequities, and social exclusion have been eliminated, but that the categories of exclusion have changed to become more technological. Now certain groups are even more marginalized and inequities are more dramatic as a result of the various sociological factors that determine access and control. This "Fourth World" (Castells, 1998) represents severely disadvantaged black holes of inequity across the globe that have increasingly little chance of reclamation. They are open to exploitation by the negative aspects of globalization but have no technology to take advantage of any counteracting, potentially positive democratic opportunities.

Another perception of globalization is that it is the next stage in a process of exploitation that began in the colonial period, in which the dominating power of the formerly colonial nations such as Britain, France, and Spain are replaced by powerful global corporations and in turn supported by international agencies such as the International Monetary Fund and the World Trade Organization (Raghavan, 1997). It is interesting to speculate about the form of globalization after the stabilization of the international financial crisis, which began in 2008. Just as corporations replaced colonial powers, what might replace the corporations?

Other outcomes of globalization include the spread of liberal democracy, the decline of authoritarian regimes, and a developing interconnectedness of the global community. Friedman (2006) calls this the "globalization of the local" (p. 505). "Indeed it is becoming clear that the flat-world platform, while it has the potential to homogenize cultures, also has, I would argue, an even greater potential to nourish diversity to a degree that the world has never seen before" (p. 506).

Globalization is not just about the economics of labor, capital, and market forces. It has also resulted in new forms of knowledge, new modes of communication, new ways of sharing work, and alternative forms of entertainment. These new tools possess a potential power equivalent to that of global corporations and international organizations. It therefore seems reasonable to utilize a postmodern perspective in which understanding is developed by viewing society in terms of knowledge power (Scholte, 2005), which is, likewise, a discourse about democracy when that is perceived as a function of power relationships.

In this sense, Dirlik (1994) characterizes postcolonialism as a child of postmodernism. For example, Foucault (1970), as a proponent of postmodernism, suggests that each historical era is characterized by a particular form of knowledge. Postmodernists attribute a form of rationalism as the dominant knowledge framework in society, emphasizing the subordination of nature to human control, objectivist science, and instrumentalist efficiency. The valuing of such a discourse ferments societies wherein economic growth, technological control, and bureaucratic surveillance provide the basis for globalization. An aspect of globalization then becomes the imposition of Western rationalism on all cultures.

Scholte (2005, p. 150) describes this rationalism as having four main features through which it promotes globalization. First, it is secularist and does not acknowledge transcendent and divine forces. Second, it is anthropocentric and seeks to understand reality in terms of human interests. Third, it has a science focus, which understands the world through incontrovertible truths discoverable through the application of objective research. Finally, it is instrumentalist and values an efficient solution to immediate problems.

So rationalism, as the basis of globalization, seeks to dominate natural forces for human purposes in order to promote capitalist production and economic efficiency throughout the cycles of global development. Technology, in its broadest sense and in all its forms, has been, and continues to be, integral to the effectiveness of globalization forces. Hence, a critical understanding of the role of technology within this milieu is an essential literacy for global citizens to possess.

Technology Education

In the context of technology education, forms of rationalism could be explicated in a number of ways. During colonial times in colonized countries, the modernist approach could be characterized by the representation of technology education as modern woodwork and metalwork, regardless of significant indigenous technologies related to construction (thatch and mud brick), hunting, food preservation, or appropriate agricultural technologies. This type of rationalist approach was clearly related to notions of progress and the determination of a single path toward what was clearly a Western conception of progress, which had resulted in the superiority of the "north" (Ullrich, 1993). To this end, the imposed technologies represented antidemocratic forces and were utilized to embed power with the colonizers; the form of technological literacy was entirely inappropriate and had the effect of inhibiting democratic development rather than promoting it.

The emergence of rationalist knowledge as an aspect of globalization (Castells, 1998) clashes with the developing postmodern notion of cultural respect and regional independence. As a counterpoint to this force, Van Wyk

(2002) proposes indigenous (technological) knowledge systems (IKS) as a framework within which diverse learners may construct knowledge from multiple perspectives that are meaningful to them. Van Wyk presents IKS as a critical framework, rather than a term with definitive meaning, that seeks to be inclusive and transformative. It is significant that this way of thinking about knowledge frameworks emanates from a South Africa that is struggling to develop relevant and democratic forms of knowledge in response to many years in which technologies (and technology education) were weapons in the armory of totalitarian apartheid. This focus is supported by Keirl (2003) in his call for technology education to adopt a critical and creational approach to knowledge development, placing students at the center of learning and so providing the opportunity to refute what is perceived to be the undesirable aspects of globalization.

A continuing phenomenon that may seem inconsistent in a postmodern international education environment is the existence of international curriculum organizations, which, by their very role, imply that there is a universal curriculum applicable to all regardless of national or regional culture or history. International curricula are in some ways the educational equivalent of multinational globalization through their ignorance of the local and the homogenization of cultural values.

As Schostak (2000, p. 48) argued, "there can be no grand narrative concerning what is good for all. Standardization to create the curriculum is patently absurd in a context of change that is so fast, so diverse, and so technologically and culturally creative." A global curriculum would seem to align more with a colonial than a postmodern environment through the promotion of totalizing forms of Western knowledge. Even worse (author's bias) is that the recipients pay a significant amount of money for the curriculum, often from a national or school budget, which is invariably stretched. Those who can afford the significant costs of adopting an internationally recognized curriculum are often those who least need it as a tool of development and an entrée into international educational equivalence.

However, the adoption of an international curriculum is rapidly increasing around the globe. The International Baccalaureate is expanding at the rate of about 14 percent annually. The main reasons for this are directly related to globalization. The forces of globalization encourage the acquisition of educational qualifications that are internationally acceptable. Allied with this is a developing mistrust of locally developed educational curricula, particularly in the United States. These forces conspire against both a postmodern critique of global developments and a valuing of local culture.

Recent Trends in Technology Education

Very few of the current national and international grand themes of technology education are indisputably successful. A focus on the development of technological literacy is probably the most widely touted broad goal of technology education. A part of the rationale for the goal of technological literacy is an attempt to attract equitable treatment for technology education along-side the other school literacies of reading, writing, and mathematics (See-mann &Talbot, 1995; Williams, 2005). Very little progress has been made in moving toward this equity of learning areas. This is reflected, for example, in the introduction of national testing in literacy and numeracy in Australia (in a state-based educational system) and the No Child Left Behind Act in the United States, which also focused on reading and math. Even in the United Kingdom, rather than focus on technology as a literacy of equal importance, research was conducted on how the study of technology could enhance the core curriculum elements of literacy and numeracy (Stables, Rogers, Kelly, &Fokias, 2001), implicitly relegating the place of technology as a means to another end rather than an end in itself.

To a certain extent, the United States has led the way in articulating the goal of technological literacy with the development of the Standards for Technological Literacy (International Technology Education Association, 2000) and associated publications. However, in the United States, a reason-able time frame to permit the implementation of these standards has been usurped by a change in the direction of educational policy having a focus on engineering as the organizational structure for technology education to the extent that the International Technology Education Association is now contributing to the development of K–12 engineering standards, and many universities, secondary schools, and some elementary schools are implement-ing engineering programs.

South Africa continues to develop a national technology education cur-riculum with a social reconstructivist approach. It has technological literacy as its overarching goal for students and contains some innovative and unique elements such as "appropriate indigenous technologies" as one of the con-tent areas. However, the legacy of apartheid policies of South Africa is that many rural schools have very few resources, and it will take many years before all schools enjoy a basic level of technological resources and equipment.

Similarly, China and India are in the process of developing, trialing, and implementing a national technology education curriculum. The challenges facing the curriculum developers in these countries are enormous. Not only is there no widespread educational history of technology as a school learning area, and therefore no school infrastructure, equipment, teachers, or support material, but with more than one billion people in each country, widespread change takes place very slowly.

England has a multifaceted approach toward design and technology with food, ICT, CAD-CAM, and electronics all receiving resources and teacher support material. Recently, a number of reports (Design Council, 2006) and some research have had a focus on design and creativity, a duality that is vital given the largely instrumental focus of support for design education. The stimulus for the focus on design has been an attempt to ensure the perfor-mance of business and industry rather than a concern for the development of individual students, hence the significance of the accompanying creativity, which does focus on the individual. Despite this comprehensive approach to development, the extent to which design and technology is a compulsory subject within the curriculum is being disputed.

New Zealand has recently completed a curriculum review in all learning areas. In technology, after the first formal curriculum in 1995, the review focused on moving away from a conception of technological literacy that is embedded in practice to one that is equally focused on understanding the philosophy of technology and developing technological knowledge. The two proposed new strands, "Nature of Technology" and "Technological Practice," replace the former strands, "Technological Capability" and "Technology and Society," with the "Technological Knowledge" strand remaining.

In Australia, the birth of contemporary technology education can be traced to the nationally agreed declaration of eight essential learning areas in 1989, one of which was technology. Since then, all the educationally independent states have developed technology curricula and consequently undergone curriculum revisions. But the focus of development from 2010 has been on a national curriculum. In the approach adopted, design and technology has been combined with communication technology to form a subject—the technologies—despite evidence that instances of this combination elsewhere have been fraught.

So it seems that few national or state approaches to technology education are sustainably successful. This may essentially be the history of technology education, based as it is on the shifting sands of technology. Technological literacy as the goal of technology education has appeal because it is variable, individual, and multidimensional—it can be related to national economic performance of a literate workforce, it relates to an individual's level of literacy with the implicit assumption that this will be more personally satisfying, and it can be used to relate to social responsibility in the context of a technological society. Notions of technological literacy contain the elements that can place literate individuals within the forces of globalization in such a way that they can critically exercise their democratic rights.

DIGITAL COMMUNICATION TECHNOLOGIES

An additional, more personal critique of technology education comes to mind in the context of a generation of young people who are growing up with a familiarity of communication technology that was unfamiliar to previous generations. In an attempt to classify them, those born in the 1980s and 1990s have been labeled "Generation Y." Of course, such labels are fraught, and there are many variations in terminology and parameters in an attempt to demarcate such demographics, including the Net Generation, Millennials, the iGeneration, Digital Natives, and MyPods (a fusion of MySpace and iPod).

For the purposes of this discussion, a number of the characteristics of this generation are relevant; it includes a group of students for whom technology education curricula are being designed by a different generation. Some see the current school-age Digital Natives as the representation of a generation gap between their teachers. Here, the relevant question becomes "how do we empower and protect our students in an environment that increasingly excludes us?" (McLester, 2007, p. 18). The existence of this gap is reinforced in the report *The Digital Disconnect: The Widening Gap between Internet-Savvy Students and Their Schools* (Arafeh & Levin, 2003), in which students reported a significant disconnect between their use of digital media in and out of school.

In the last few years, a new set of values has arisen, confirming the barrier between generations. This current school generation was the first to witness and use a broad range of technologies from an early age: the Internet and broadband, digital cable, cell phones, HDTV, digital cameras, digital pets, camera phones, social networking, GPS technology, online gaming, and touch screens. Of course, other generations use these technologies, but they did not grow up with them and integrate them into their lifestyle to such an extent.

Until these developments, much modern technology mitigated the opportunity to democratically engage in civic life and its associated politics (and of course still does): shopping malls, spreading suburbia, automobilization, and a vast array of home entertainment options, for example. While they may continue to work against the democratic sharing of power, they do not preclude it. And in the face of a more powerful range of technologies, they exercise diminishing influence. This range of technologies has been dubbed Web 2.0, the "people version" of the World Wide Web.

The initial promotion of the term Web 2.0 is generally ascribed to O'Reilly (2006) and refers to a perceived second generation of web-based services that are characterized by open communication, decentralization of authority, freedom to share and reuse, users' ownership of data, and an effectiveness that develops as more people use them.

Individuals both learn from and contribute to Web 2.0, whereas Web 1.0 was a learning tool—only the experts could contribute. Anyone can now review books on Amazon.com; blog about political candidates and have an effect; podcast about things they are interested in; participate interactively in car design competitions; make movies on their phones and receive 100,000 views within a week; film and distribute footage of everything from a celebrity to an abuse of power; and submit research to the world's largest encyclopedia without having to worry about peer reviews. Individuals have seized the reins of the global media and founded the new digital democracy, beating the professionals at their own game (Grossman, 2006).

The means of Web 2.0 involvement include LiveJournal, MySpace, Wikipedia, LastFM, Netflix, Facebook, del.icio.us, Flickr, and outside.in. YouTube is probably the most significant means, and it is not just one medium, but several in one. As Poniewozik (2006) explains, it is the following:

- *A surveillance system.* Millions of people, through their mobile phones, have the power to quickly and easily send any happenstance image around the world, from the London train bombing to celebrities in unguarded moments to politicians in compromising positions.
- *A spotlight.* Users have the capacity to find significance in events that the mainstream media may ignore. Programs and advertisements made for

television have been rejected and then reborn after being uploaded to YouTube.

- *A microscope.* While television news is constrained by budgets and time, this is not so for YouTube. Extreme closeup video diaries from Iraq, Israel, and Lebanon convey the confusion, humanity, and reality of war zones.
- *A soapbox.* Anyone's ideas can spread instantly, cheaply, democratically, and anarchically.

DIGITAL DEMOCRACY

The relationship between technology and democracy has varied over time. When technology first developed, accompanying the genesis of civilization as we know it, simple technologies enabled the development of communal democracy. The use of tools permitted challenges to existing power structures by enabling those with the technology to exercise their newfound power. Later, increased mobility extended relationships once limited to the family and the tribe into broader communities. This geographic extension of influence spread democratization influences by further challenging power structures.

Throughout different periods of time, technology both has facilitated the development of democracy and has been an impediment to democracy, and this equally applies today. It could be argued, as Sclove (1995) does, that the power vested in the technological elite by the technologies that they control, such as gasoline-powered automobiles or proprietary Internet tools, undermines any grassroots democratic movements that may develop in opposition to their interests. Such technologies inhibit participation in significant decision making and define social orders that constrain self-actualization and support illegitimate hierarchies. On the other hand, some technologies constitute democratic forms of power sharing, such as broadly accessible web-based technologies with international and pervasive distribution networks, which enable individuals and groups to have a voice and for their voices to count.

Information and social media technologies are an integral aspect of the broader notion of technology, and so their use and understanding must be developed as part of this literacy, which people need in order to exercise their democratic rights and comprehend the nature of global development. The extent to which this literacy is informed determines the dispositions (Williams, 2011) of individuals reacting to this development.

A range of influential information technologists from Negroponte to Gates are predicting that in the near future, access to media will become increasingly personalized. They predict a "Daily Me" newspaper will be delivered that is personally tailored to each individual, only dealing with categories of news and information that have been predetermined by the individual. Similarly, "TV Me" will just show those programs that reflect individual interests.

So there is some danger in promoting Web 2.0 as the means of democratic engagement. Williams (2006) warns that participation in Web 2.0 can be simply a celebration of self, a narcissistic infatuation. It is now possible to go about your day and consume only what you wish to see and hear:

"television networks that already agree with your views, iPods that only play music you know you like, Internet programs ready to filter out all but the news you want to hear" (2006). The problem with this is that there is a lot of information that individuals in an informed democracy need to know, with the consequent danger that "we miss the next great book or the next great idea, or that we fail to meet the next great challenge . . . because we are too busy celebrating ourselves and listening to the same tune we already know by heart" (2006).

A possible corollary of these developments is that personal cognitive systems will become unable to pursue the option of evaluating a range of knowledge and information or make selections and judgments according to certain criteria, which are core skills for democratic citizenship (Sunstein, 2007). The assumption here is that either individuals actually have the resources and power to decide and then access what it is that they need to know, or whether it is global media organizations that make those kinds of important decisions. In the latter scenario, in which the quality of "newsworthy" is ascribed by media organizations, the individual power to bypass these business decisions and access information that "feels" relevant and important through the Web 2.0 system of opportunities would place the individual in a powerful, self-determining position, enabling him or her to break away from the media's tendency to allow people to be only consumers rather than citizens.

Generally, the accumulation of power is accompanied by a level of responsibility, and unless the responsibility is felt and active, the possession of power wanes. For example, in the influential areas of politics, education, or the media, irresponsibility results in implications at the ballot box, in tenure, or in ratings. However, in the case of the power that accompanies participation in Web 2.0, responsibility can be abused without the loss of power. Fiction can parade as fact, respect for others is not an assumption, character assassinations are tolerated, and the unimportant is promoted. Developing fundamentalist groups from neo-Nazis to al-Qaeda provide all too imminent evidence of the negative aspects of the democratization of communication technologies.

The global counterpoints proposed by Barber in 1992, prior to the recognition of Web 2.0, provide an exemplar of opposing tendencies related to democracy, power, and responsibility. In "Jihad vs McWorld," he explains these forces as having "equal strength in opposite directions, one driven by parochial hatreds, the other by universalizing markets, the one re-creating ancient subnational and ethnic borders from within, the other making national borders porous from without. They have one thing in common: neither offers much hope to citizens looking for practical ways to govern themselves democratically." Both these forces have found an affinity in Web 2.0: a clear indication that its use is diverse and that an ethically based technological literacy is an important prerequisite for participation.

McClintock (2004) uses an urban political metaphor to argue that the web enables a civilized, sophisticated style of discourse akin to that identified in real (nondigital) metropolitan democratic communities. Joint (2005) points out that there is an underside to urban politics that thrives just as much virtually as it does in reality. The networks provide the opportunities for the development of squalid cyberghettos: "As a result a bastardized form of eLiteracy enables the internet thug to search, locate and colonize these spaces, while expertly circumnavigating both technical filters and moral challenges The most appalling instances of this depraved web virtuosity are internet videos of terrorist slayings, expertly spread across the net in an attempt to amplify the impact of political murder on democratic electorates" (2005, p. 81).

The nature of the democracy that is developed through virtual participation is discussed by Barber (2004) in his book *Strong Democracy*, which he characterizes as not necessarily always being direct, but always incorporating strong participatory and deliberative elements. Deliberate participation through social media has been evident recently in the popular uprisings that have occurred against a number of regimes across the world. Whether the outcomes constitute strong democracy or are more satisfactory than their precursors remains for history to decide, but there is little doubt that social media technology played a significant role in facilitating individuals' capacities to exercise what they perceived as their democratic rights.

Aligned with this notion of national development is a dilemma for many countries in which there is a political and economic desire to move ahead technologically, while at the same time centralizing and controlling information. China is probably the most obvious example of this dilemma, but when applied to Cuba, Venegas (2003, p. 192) terms it "the dictators digital dilemma."

THE GROUNDING OF THE RELATIONSHIP

So, given that entrée into this electronic participatory democracy is freely available and that everyone's right to speak can become a reality, there also exists the converse reality that this is disassociated from the responsibility to critically listen to what arises from the more open Web 2.0 sphere. There is no system of checks and balances on the power that is derived from participation, and so a question arises as to the development of responsibility.

A technology education approach that addresses the development of technological responsibility through a democratic, information-technology medium (Web 2.0) would be an appropriate dimension for technology education to address. It is not a radical departure from the goal of technological literacy or creative and innovative design but more a recognition of the dynamism of both the process and the content of technology. The increase in the power of the individual that accompanies Web 2.0 and the consequent increase in responsibility within a participatory democracy provide a context for a renewed focus in technology education on the social and ethical nature of technology.

A postmodern and enquiry-based narrative of technology, which pays mutual attention to both the local context and impinging global developments, would seem to provide an appropriate framework for the development of this type of technology education. McNeil's (1981) conceptualization of a curriculum of personal relevance resonates with this personal preparation for participation in a digital democracy. The goal of such a curriculum is to produce a self-actualizing, autonomous, authentic, healthy, happy person (Petrina, 1992) through an integrated focus on the cognitive, affective, and psychomotor areas of development. Elements of McNeil's proposal include the following:

1. All individuals participate in the curricular and learning processes.
2. There is integration of the material being learned and integration in the humanistic approach taken.
3. The subject matter is emotionally and intellectually relevant to the participants.
4. The person is the object of the learning.
5. The goal is to develop the whole person within a social/technological context.

A humanistic approach to technology is appropriate in this context given its integrative potential and the individually autonomous balance it provides to the potentially deterministic forces of technology. Children's feelings and morality about technology have a significantly higher level of potency when they are channeled through the mechanisms that have become Web 2.0.

Specific teaching techniques are implied in this approach to curriculum—those that encourage planning and spontaneity, expression, insight, and reflective thought. Students can access data from a broad range of sources and work in teams to collaboratively produce at a distance. The notion that education only occurs at school can be broadened in the sense that students can engage in design and technology processes from a broad range of locations. The psychology of "thinking and doing" as the mode of learning in technology education resonates with the interactivity of Web 2.0. The web-based resources available provide stimulus for both developing ideas and doing something with those ideas.

Relevant content knowledge in technology is defined as that which is needed in order to progress a student's thinking and action toward resolution of a problem. The source of content then is determined by its relevance to the individual student's environment and their immediate concerns. In the context of Web 2.0, students are able to locate, organize, and evaluate information from a broad range of sources and so are more likely to access information appropriate to a specific task.

One of the important roles of technology education in this context is to harness access to diverse positions and fight against what Gozálvez (2011) terms "multiple digital inbreeding," or the tendency to communicate only with known groups that are relatively stable in the Internet. These groups are comfortable because they orient around certain beliefs or ideological perspectives but neglect broader, more common interests, which are at the core of the democratic state.

The basis of the learning process is more amenable to authentic problems and significant questions due to the ease of research and communication available. The collection of relevant data and the use of multiple processes to explore alternatives are readily available to students. Throughout the processes of designing and making, opportunities are presented to focus on responsible-citizenship aspects of technological literacy through the safe and responsible use of information and a positive attitude toward technologies that support collaboration and learning.

So this scenario involves utilizing the technologies of Web 2.0 as another tool in designing and making processes that make relevant and responsible technological literacy the ultimate goal of technology education. Literacy in this context means attaining a form of fully developed digital citizenship, one predicated on the possession of an ability to interpret, navigate, and shape the landscape of virtual democracy (Joint, 2005, p. 81).

CONCLUSION

Technology is of course not the only factor determining the efficacy of a participatory democracy, through either impairing it or facilitating it; there are many other factors well beyond the influence of schooling. Elements of the argument of this chapter, that one focus of technology education in a postmodern age should include preparation for engagement in a participatory democracy, may be contestable. The fact remains, though, that Web 2.0 enables a level of individual participation heretofore unavailable and provides a medium of personal relevance to students.

Many technologies are the substance of public controversy and increasingly consume media attention and political preoccupation. Consider power generation, the causes of what used to be natural disasters, telecommunications and security systems, defense and weapons systems, hazardous waste disposal, and resource exploration. Powerful political actors dominate the discussion and control of significant technological issues—politicians, government administrators, corporate leaders, and representatives of special interest groups. This current system of discourse and decision making is inadequate because it

- excludes lay citizens from anything but a trivial role
- often raises questions after many of the most important decisions have already been made
- evaluates technology on a case-by-case basis rather than having a philosophical starting point as a basis for decision making
- focuses on a few high profile technologies at the expense of the mass of emerging and existing technologies
- directs attention to material issues rather than the often more important social and cultural issues
- excludes discussion of the influence of technology on democracy (Sclove, 1995, p. 240)

Left to current patterns of control and decision making, the world is not going to become a better place in which to live for all its inhabitants. But this degeneration is insidious; standards of democracy, personal freedom, and well-being are diminished in small steps, which neither make big news nor become the focus of attention. So life progressively becomes more technologically determined (Ellul, 1964) without attracting attention to the trend.

The thesis of this chapter is that a technological literacy that aspires to a form of fully developed digital citizenship, one predicated on the possession of an ability to interpret, critique, navigate, and shape the landscape of virtual democracy, facilitates a move toward a more democratic technological order. This element of technological literacy involves not only using the web in a passive, albeit intelligent, way but also knowing how to mold it to reflect the values that we think it should have.

REFERENCES

Arafeh, S., & Levin, D. (2003). The digital disconnect: The widening gap between Internet-savvy students and their schools. In C. Crawford et al. (Eds.), *Proceedings of society for information technology and teacher education international conference 2003* (pp. 1002–1007). Chesapeake, VA: American Association of Clinical Endocrinologists.

Barber, B. (1992). Jihad vs. McWorld. *The Atlantic, 269*(3).

Barber, B. (2004). *Strong democracy: Participatory politics for a new age.* San Francisco, CA: University of California Press.

Bernstein, H. (2000). Colonialism, capitalism, development. In A. Thomas (Ed.), *Poverty and development into the 21st century* (pp. 241–270). Milton Keynes, UK: Oxford University Press.

Boyne, R., & Rattansi, A. (1990). The theory and politics of postmodernism: By way of an introduction. In A. Rattansi (Ed.), *An introduction to post modernism and society.* London, UK: Macmillan.

Bridgstock, M., Burch, D., Forge, J., Laurent, D., & Lowe, I. (1998). *Science, technology and society.* Cambridge, UK: Cambridge University Press.

Cambridge International Examinations (CIE). Retrieved August 28, 2006, from http://www.cie.org.uk.

Castells, M. (1998). *End of millennium.* London, UK: Blackwell.

Design and Technology Association. (1999). *Cross-curricular links within the primary curriculum.* Wellesbourne, UK: Author.

Design Council. (2006). *Design council annual 2005/06.* London, UK: Author.

Dirlik, A. (1994). The postcolonial aura: Third world criticism in the age of global capitalism. In P. Mongia (Ed.) *Contemporary postcolonial theory* (pp. 294–319). London, UK: Arnold.

Dirlik, A. (1997). *The postcolonial aura: Third world criticism in the age of global capitalism.* Boulder, CO: Westview.

Ellul, J. (1964). *The technological society.* Trans. John Wilkinson. New York, NY: Knopf.

Foucault, M. (1970). *The order of things.* London, UK: Tavistock.

Friedman, T. (2006). *The world is flat.* London, UK: Penguin.

Gandhi, L. (1998). *Postcolonial theory.* Sydney, Australia: Allen and Unwin.

Gozálvez, V. (2011). Education for democratic citizenship in a digital culture. *Scientific journal of media literacy* (36), 131–138.

Grossman, L. (2006). Time's person of the year: You. *Time*. Retrieved January 24, 2007, from http://www.time.com/time/magazine/0,9263,7601061225,00.html.

Hales, J., & Snyder, J. (1981). *Jackson's Mill industrial arts curriculum theory*. Charleston, NC: West Virginia Department of Education.

Harrison, P. (1983). *The third world tomorrow*. Hammondsworth, UK: Penguin.

International Baccalaureate Organization (IBO). Retrieved August 30, 2006, from http://www.ibo.org.

International Labour Organization (ILO). (1977). *Matching employment opportunities and expectations: A program of action for Ceylon*. Geneva: United Nations.

International Technology Education Association (ITEA). (2000). *Standards for technological literacy*. Reston, VA: Author.

Joint, N. (2005). Democracy, eLiteracy and the Internet. *Library Review, 54*(2), 80–85.

Keirl, S. (2003). Globalisation on the go: Implications for design and technology education. In E. Norman and D. Spendlove (Eds.), *DATA international research conference proceedings* (pp. 57–62). Wellesbourne, UK: Design and Technology Association.

Klages, M. (1997). *Postmodernism*. Retrieved March15, 2005, from http://www.colorado.edu/English/ENGL2012Klages/pomo.html.

Loomba, A. (1998). *Colonialism/postcolonialism*. New York, NY: Routledge.

Marsden, K. (1973). Progressive technologies for developing countries. In R. Jolly, E. de Kadt, H. Singer, & F. Wilson (Eds.), *Third world employment*. Hammondsworth, UK: Penguin.

McClintock, R. (2004, June). *eLiteracy and city civics*. Keynote presented at the eLit2004 Conference, St. John's University, New York.

McLester, S. (2007). Technology literacy and the MySpace generation. *Technology and Learning, 27*(8), 16–20.

McNeil, J. D. (1981). *Curriculum: A comprehensive introduction* (2nd ed.). Boston, MA: Little, Brown and Company.

O'Reilly, T. (2006). Web 2.0 compact definition. Retrieved February 18, 2007, from http://radar.oreilly.com/archives/2006/12/web_20_compact.html.

Petrina, S. (1992). Curriculum change in technology education: A theoretical perspective on personal relevance curriculum designs. *Journal of Technology Education, 3*(2).

Poniewozik, J. (2006). The beast with a billion eyes; on the web, anyone with a digital camera has the power to change history. *Time*. Retrieved March 20, 2014, from content.time.com/time/magazine/article/0,9171,1570808,00.html.

Raghavan, C. (1997). *WTO conference: How the developing countries lost out*. Third World Resurgence, 77–78.

Scholte, J. (2005). *Globalization*. New York, NY: Palgrave Macmillan.

Schostak, J. (2000). Developing under developing circumstances: The personal and social development and students and the process of schooling. In H. Altrichter and J. Elliot (Eds.), *Images of educational change* (pp. 37–52). Philadelphia, PA: Open University Press.

Schumacher, E. (1973). *Small is beautiful*. London, UK: Blond and Briggs.

Sclove, R. (1995). *Democracy and technology*. New York, NY: Guilford Press.

Seemann, K., & Talbot, R. (1995). Technacy: Towards a holistic understanding of technology teaching and learning among Aboriginal Australians. *Prospect, UNESCO Quarterly Review of Comparative Education, 25*(4), 761–775.

Stables, K., Rogers, M., Kelly, C., & Folias, F. (2001). *Enriching literacy through design and technology evaluation project*. London, UK: Goldsmiths College, University of London.

Sunstein, C. (2007). *Republic.com 2.0*. Princeton, NJ: Princeton University Press.

Ullrich, O. (1993). Technology. In W. Sachs (Ed.), *The development dictionary: A guide to knowledge as power* (pp. 275–287). Johannesburg: Witwatersrand University Press.

United Nations (UN). (1995). *States of disarray: The social effects of globalization*. Geneva: United Nations Research Institute for Social Development.

Vaitsos, C. (1973). Patents revisited: Their function in developing countries. In C. Cooper (Ed.), *Science technology and development* (pp. 71–98). London, UK: Frank Cass and Co.

Van Wyk, J. (2002). Indigenous knowledge systems: Implications for natural science and technology teaching and learning. *South African Journal of Education, 22*(4).

Venegas, P. (2003). Will the Internet spoil Fidel Castro's Cuba? In H. Jenkins, and D. Thornburn (Eds.), *Democracy and the new media* (pp. 179–202). Cambridge, MA: MIT Press.

Williams, B. (2006). But enough about you . . . *Time*. Retrieved January 29, 2007, from http://www.time.com/time/magazine/article/0,9171,1570707,00.html.

Williams, P.J. (2005, April). Technology education in Australia: Twenty years in retrospect. Paper presented at the PATT 15 Conference, The Netherlands.

Williams, P.J. (2011). Dispositions as explicit learning goals for engineering and technology education. In M. Barak, & M. Hacker (Eds.), *Fostering human development through engineering and technology education* (pp. 89–102). Rotterdam: Sense.

GENDER AND TECHNOLOGY

CHAPTER 5

REENVISIONING OUR KNOWLEDGE TRADITION

FROM GENDERBLIND TO GENDERAWARE

Mary Kirk

INTRODUCTION

The ways in which we have learned to think about technoscience[1] are influenced by our assumptions about gender—our notions of both maleness/masculinity and femaleness/femininity. However, this influence is often invisible because most view technoscience as genderneutral when it is actually just genderblind (Wajcman, 2004). Technoscience has been defined as a male/masculine domain that excludes the female/feminine (sometimes literally; Frehill, 2004; Lohan & Faulkner, 2004; Millar, 1998). This genderblindness has not only excluded multiple perspectives as technoscientific creators; it has also limited the creations themselves.

Both the technologies we create and the ways in which those technologies are used are profoundly influenced by gender-blind assumptions. For example, Judy Wajcman (2004) tells a tale of two Internets. One reflects the liberatory potential of undermining old social relations, redefining gender roles, strengthening participatory democracy via electronic communities, reaching across our cultural divides, and sharing ideas beyond the control of any one group. The other reflects the oppressive limitations of eliminating personal privacy and expanding the white male hacker culture, burgeoning international racism, economic colonialism, the dominance of ideas from English-speaking cultures, and the massive explosion of pornography and the international sex trade. The question of which Internet predominates will be answered by those who shape its continued creation and by whether they have become genderaware.

Wajcman (2004) suggested adopting a technofeminist perspective to help us become genderaware and to see the dynamic "relationship between gender and technology, in which technology is both a source and a consequence of gender relations" (p. 7). Harding (2008)—who engaged in a rich interdisciplinary analysis that included feminist science studies, mainstream "traditional" science and technology studies, modernism, and postcolonial studies—recommended that our perspective extend beyond the exclusive concerns of the Northern Hemisphere. Wajcman (2004) underlines the need for this broader vision via the following example of how North, South, East, and West are already interconnected in complex ways with varying economic and cultural benefits/costs:

> For a young woman in the West, her silver cell phone is experienced as a liberating extension of her body. The social relations of production that underpin its existence are invisible to her . . . [Cell phones] require the scarce mineral Coltan. One of the few places where this can be found is Central Africa, where it is mined under semi-feudal and colonial labour relations, to provide a raw product for Western multinational companies. The sharp rise in the price of Coltan on global markets has local effects, accentuating exploitation and conflict among competing militias, with the very specific consequences for women that military conflict brings—namely, rape and prostitution. (pp. 121–122)

Although wars in Coltan-producing states such as Rwanda are fueled by global economic concerns, they are often described to the world as ethnic conflicts. Meanwhile, such wars have led to food crises that have raised serious environmental concerns as gorilla populations are decimated to sell "bush" meat to the hungry. As this example shows, the interactions between people and technoscience are dynamic and multidimensional. And that is actually the good news.

From a gender-aware perspective, we have the opportunity to include a multiplicity of viewpoints in developing technologies that serve human beings instead of putting human beings in service to them. Technoscience has the potential to be a site for redefining culture in service of our highest human good or for reifying our current gendered power relations—which potential arises will be the direct result of our degree of genderawareness.

This three-part chapter explains how our knowledge tradition teaches genderblindness: "Dualistic Delusions of Gender"[2] describes how we learn gender and explores the nature/nurture debate; "Gendered Philosophy of Science (and Technology)" reveals the ways in which technoscience is gendered, while claiming to be genderneutral; and "Invisible Women in Science and Technology" shares the growing history of women in science and technology, which (until recently) has been omitted from our knowledge tradition.[3] By exposing our learned genderblindness, I hope to help readers develop the genderawareness that is critical to increasing technological literacy in the twenty-first century.

Dualistic Delusions of Gender

As I discussed in *Gender and Information Technology: Moving beyond Access to Co-Create Global Partnership* (2009), our dualistic obsession with *gender*—the characteristics of behavior associated with female and male—deludes us into attempting to categorize human identities, unique experiences, and complex ideas into simplistic either/ors. Table 5.1 lists a few typical gender-assigned characteristics that tend to be true across cultures.

We learn to stay within and assert the boundaries of these "gender boxes" from all our social institutions, such as family, education, media, government, and religion. You can easily identify the boundaries of the boxes in your culture by noticing when women or men are chastised for their "inappropriate" behavior (Kirk, 2009).

Many have debated whether women and men differ due to nature (biology, sex, genes) or nurture (environmental influences). Women's increasing entry into educational and professional fields previously dominated by men has not ended this debate. When even our best scientific knowledge recognizes dynamic and varied influences on nature and human development, why do we still debate nature/nurture? For example, Nobel Prize–winning physicist Werner Heisenberg founded quantum mechanics on the principle that observing an object can influence its behavior. Developmental psychologist Urie Bronfenbrenner recognized that human development involved dynamic relationships between individuals and their environments, saying that "individuals influence the people and institutions of their ecology as much as they are influenced by them" (Lerner, 2005, p. ix). More recently, physicist and feminist science studies scholar Evelyn Fox Keller (2010) described what the new science of genetics has taught us: "What we have learned has not so much answered earlier questions as it has transformed them. We have learned, for instance that the causal interactions between DNA, proteins, and trait development are so entangled, so dynamic, and so dependent on context that the very question of what genes do no longer makes much sense. Indeed, biologists are no longer confident that it is possible to provide an unambiguous answer to the question of what a gene is" (p. 50).

Table 5.1 The Gender Boxes	
Male	Female
Science	Nature
Hard	Soft
Strong	Weak
Rational	Emotional
Active	Passive
Assertive	Submissive
Competitive	Cooperative

When even science recognizes the dynamic influences *between* nature and nurture, why are we still asking the question? In *The Mirage of a Space between Nature and Nurture*, Keller (2010) explored how the notion that nature and nurture are separable came to be taken for granted, how muddled our language around these questions has been, and where today's genetic discoveries have led. Keller credits Francis Galton (Charles Darwin's cousin) with the "first explicit use of the terms nature and nurture as an unambiguous disjunction" (p. 20). Interestingly, Galton's ideas on "hereditary improvement" ultimately led him to coin the term "eugenics" for a strategy of race improvement (p. 27)—a clear example of the ways in which science has been, and can continue to be, influenced by the culture from which it arises.

I recommend a cautious stance toward research that focuses on difference due to (1) the ways in which it has been used to discriminate against people by broad categories of identity; (2) the ways in which it seeks to separate and divide rather than unite us; and (3) the poor quality of some of this research. For an example of my first and second points, I point to the so-called science of eugenics. My third point continues to be well addressed by feminist scholars who have been doggedly challenging the poor science in some of this "difference" research for more than twenty years, beginning with Ruth Bleier's (1991) analysis of inaccurate and misleading sex differences research to Cordelia Fine's (2010) more recent analysis of questionable interpretations of brain scans conducted by neuroscientists.

The nature/nurture question is predicated on the false assumption that nature and nurture are separable. Current scientific knowledge belies this premise. Keller (2010) proposed a new question: "Let us ask not how much of any given difference between groups is due to genetics and how much to environment, but rather how malleable individual human development is, and at what developmental age . . . there is no reason to privilege birth as a cutoff point—development is lifelong, and so it its plasticity" (p. 84). Keller's new question comes closer to acknowledging the reality that gender identity is not static but constantly evolves in response to the environment.

Gendered Philosophy of Science (and Technology)

The rigidity and persistence of dualistic thinking about gender is also reflected in the equation of science/technology with maleness and nature with femaleness. Francis Bacon (1561–1626) and Rene Descartes (1596–1650) are considered the fathers of scientific reasoning, mathematics, and modern philosophy. Feminist philosophers and feminist science studies scholars described how their early writings also helped reify the definition of science as male and nature as female (Bleier, 1991; Bordo, 1986; Keller, 1985; Merchant, 1980; Schiebinger, 1993; Schiebinger, 1999).

Carolyn Merchant (1980) shared many examples of the implicit power dynamics of science (male) over nature (female) reflected by these early thinkers. These ideas may not have led to the gendered split of science over nature, but they arose concurrent with a "new concept of the self as a rational master of the passions housed in a machinelike body," which began to "replace the concept of the self as an integral part of a close-knit harmony of organic parts united to the cosmos and society" (p. 214). Ultimately, three concepts—mechanistic thought, order, and power—deeply informed Western politics, religion, and science and contributed to the development of science as a domain that increasingly excluded the female (and women themselves).

Susan Bordo (1986) viewed Rene Descartes's seventeenth-century ideas on objectivity and scientific rationalism through the new lens of developmental psychology, specifically separation and individuation. Bordo suggested that Cartesian objectivism may reflect a sort of separation anxiety resulting from the flight from the feminine occurring as Descartes's peers moved away from the concept of "the organic female universe of the Middle Ages" (p. 442). Objectivity was to be achieved by transcending the body (nature and the female) and relying on reason alone, which Bordo claimed resulted in a masculinization of thought.

Ruth Bleier (1991) also examined the ways in which seventeenth-century Baconian dualism "elaborated the metaphors of science in sexual and gendered terms, with science as male and nature as female, a mystery to be unveiled and penetrated" (p. 6). Woman was embodied in "the natural, the disordered, the emotional, the irrational," and man "epitomized objectivity, rationality, culture, and control" (p. 6). Note the similarity to the contents of the gender boxes I outlined earlier.

Bleier (1991) also challenged the notion of scientific objectivity, pointing to another reason that the "difference" research I mentioned earlier is so problematic: "By draping their scientific activities in claims of neutrality, detachment, and objectivity, scientists augment the perceived importance of their views, absolve themselves of social responsibility for the applications of their work, and leave their (unconscious) minds wide open to political and cultural assumptions" (p. 29).

As long as technoscientists cloak themselves in the myth of objectivity, they remain determinedly blind to the ways in which their work, and even the questions they ask, is influenced by cultural assumptions. Figueroa and Harding (2003) similarly questioned whether science could be separated from social influences, citing W.V.O. Quine, who "proposed that scientific and everyday beliefs were linked in networks. How scientists theorized nature's order and chose to revise their hypotheses when faced with counterevidence depended in part on the ontologies, logics, and epistemologies they brought to their work, largely unconsciously, from their particular cultural contexts" (p. 2).

Feenberg (2003) offers an example of the significance of sociocultural context in the development of the computer keyboard, which appears culturally neutral to Westerners. However, Feenberg explains that if computers had "been invented and developed first in Japan, or any other country with an ideographic language, it is unlikely that keyboards would have been selected as an input device"; the early input devices would more probably have been designed "with graphical or voice inputs" (pp. 242–243). Feenberg shared another example from O-Young Lee, who "argues that the triumph of Japanese microelectronics is rooted in age-old cultural impulses . . . to miniaturize, evident in bonsai, haiku poetry, and other aspects of Japanese culture" (Feenberg, 2003, pp. 242–243).

In her 1986 review of the writings of feminist philosophers and feminist science studies scholars, Sandra Harding identified five themes: issues of equity that influence women's participation; sexist and racist scientific research; social constructivist ideas that challenge the notion of "pure science"; deconstruction of earlier writings on science, especially examining the use of language; and epistemological analyses. Now, more than twenty years later, Harding has expanded that discourse from feminist standpoint theory, to the concept of strong objectivity, and most recently to considering technoscience in a postcolonial context (Harding, 1998, 2006, 2008). Harding's (2006) work linking postcolonial studies with feminist science studies reflects the broadly shared assumption among feminist scholars that "our methodological and epistemological choices are always also ethical and political choices" (p. 156) and that "scientific standards, like the scientific claims they claim to justify, are always socially situated" (p. 145).

INVISIBLE WOMEN IN SCIENCE AND TECHNOLOGY

Women have always participated in science and technology. However, until recently, as with many other areas of our knowledge tradition, the history of science and technology was blind to women's contributions—it was largely male and pale (white) (Rossiter, 2012, p. 275). Women such as Evelyn Fox Keller (1983), Betty Toole (1992), and Autumn Stanley (1995) began to document women's histories in science. A few years later, Wini Warren (1999) recounted stories of black women in science. More recently, Thomas Misa (2010) and Janet Abbate (2012) have documented the histories of women in computing. Leading historian of science Margaret Rossiter (2012) described the event (while a graduate student at Yale in the early 1970s) that inspired her to document women's achievements:

> I asked my Yale professors of the history of science whether there had ever been any women scientists. The emphatic answer was assuredly not. (Even Madame Curie with her two Nobel prizes was discounted as a mere drudge who helped her husband Pierre.) Years later, as a postdoctoral fellow at the Charles Warren Center at Harvard University and having found several hundred women in the early editions of the American Men of Science, I pressed on with this topic, determined to write a one-volume survey of women scientists from the beginning, whenever that was, to the recent past. (p. xvi)

Rossiter's project exploded into what is now a three-volume history of women scientists in America that lays a solid foundation on which future scholars can build. Although much more is needed, Rossiter's meticulously researched work identifies structural and cultural themes that reveal patterns of women's ins and outs in technoscience in their historical context. Perhaps the greatest outcome of Rossiter's work is identifying the ways in which the historical legacy of women's participation still influences women today.

In *Women Scientists in America: Struggles and Strategies to 1940*, Rossiter (1982) chronicles the "series of limited stereotypes, double binds, resistant barriers," and other "no-win situations" that women historically faced in colleges and universities and demonstrates how historical events (such as World War II) alternatively encouraged women's participation in education and the professions and then pushed them out (p. xvii). Until the founding of a few women's colleges, such as Smith (1871), Wellesley (1875), and Bryn Mawr (1885), women only had limited access to higher education. Interestingly, these single-sex learning environments supported the success of women in science just as many contemporary scholars have described: the presence of female role models and mentors; smaller classes; allowing women to talk more; curricular content that is more "female-friendly"; and cooperative peer dynamics (Barker & Aspray, 2006; Rossiter, 1982; Sadker & Sadker, 1995; Warren, 1999).

Access to undergraduate education for African American women was even more circumscribed; segregation forced most to attend historically black colleges and universities (HBCUs). Like the women's colleges, the HBCUs were successful in producing women scientists, but their facilities were often underfunded, and professional opportunities for graduates were limited to teaching in HBCUs (Rossiter, 1982). It took many more decades for African American women to have access to doctorates in math and science, and that was also influenced by race. In 1886, Winifred Edgerton Merrill (Columbia University) became the first white woman to earn a doctorate in mathematics, but it wasn't until 1949 (63 years later) that Evelyn Boyd Granville (Yale University) became the first African American woman to earn a doctorate in mathematics (Williams, 1999). In 1973 (24 years later), Shirley Ann Jackson became the first African American woman to earn a doctorate in physics. From 1933 to 1973, these women were two of only ten doctoral firsts among African American women in science (Rossiter, 1995, p. 83). Access was not the only barrier, as Amy Bix's (2000) history of women at the Massachusetts Institute of Technology (MIT) from 1871 to 2000 shows; women experienced an inhospitable climate that ranged from marginalization to outright sexism even at this elite engineering institution.

In terms of employment, teaching was the primary career path for educated women in the 1880s, and overt gender discrimination led to a 40 percent salary gap between women and men (Rossiter, 1982, p. 5). There was a similar pattern with women scientists at universities, who often taught heavy loads as "volunteer professors" or volunteered as researchers (Rossiter, 1995, p. 141). Although the scarcity of skilled male workers during the war years created tremendous employment opportunities for women scientists, they suffered from openly accepted salary inequities. For example, the US federal government wanted women in certain positions because they *could* pay them less. In 1938, while the average salary for men in one civil service category was $3,214, women in the same category earned an average salary of $2,299—almost 40 percent less (Rossiter, 1982, p. 235). The gendered salary gap that still exists today is a legacy of this history.

In *Women Scientists in America: Forging a New World since 1972*, Rossiter (2012) explained how women scientists began to challenge salary discrimination, form their own scientific and technical organizations, and raise funds for projects that "made inroads into a sexist and elitist system," notably amid major political swings "mostly to the right" (p. xvii). She highlighted the stories of biochemist Sharon L. Johnson, anthropologist Louise Lamphere, and chemist Shyamala Rajender, whose salary discrimination law suits enforced new laws and altered public perception (pp. 31–35). The 1970s were marked by "a slow trickle of firsts," such as "the first woman hired in any science or engineering college, the first woman in a particular department, the first woman tenured, the first woman full professor, and even the first woman chair or assistant dean" (p. 26). But the demands these trailblazers faced in relation to "inspiring female graduate students" were great: "All too often they too were young, isolated, overworked, and frightened, hiding in their offices, unsure how to set up and run a laboratory, and resentful that on top of everything else they were expected to deal with the unhappy female graduate students—something they did not know how to do and were sure would get them into more trouble if they succeeded and perhaps even if they tried" (Rossiter, 2012, p. 117). These women's bold efforts toward change consumed "time, money, energy, emotion, and health that in a more perfect universe might have gone into scientific teaching or research" (Rossiter, 2012, p. 39). In fact, many women today juggle similar competing demands.

The legalized segregation that led most African American women in science to stay and teach at HBCUs resulted in more women in more fields, giving them the critical mass they needed to encourage even more women—at least within the HBCUs. This caused HBCUs to become significant contributors to the numbers of women of color in science and engineering. Although "the 103 HBCUs enrolled only 2 percent of the nation's college students in 1994, together they accounted for 28 percent of the bachelor degrees earned by African Americans," and "of the eighteen physics departments in the nation that graduated the most women with a BS in physics in 2005, seven were at HBCUs [and] . . . of the African American women who later earned PhDs in science . . . almost half were from the HBCUs, especially the two historically black women's colleges, Spelman College and Bennett College" (Rossiter, 2012, p. 56).

Between 1970 and 2000, the number of women completing degrees in science and engineering more than doubled for bachelor's degrees and quintupled for doctoral degrees. But for bachelor's degrees, the numbers "varied greatly by field" and actually collapsed in computer science (Rossiter, 2012, pp. 41–95). While recent data show a fairly steady increase in women in biology and psychology, the numbers in engineering and computing have not grown at the same pace. One reason for this discrepancy in women's participation is that biology and psychology were areas in which women had the least resistance historically. Over the 1900s, as more women entered those disciplines, they served as mentors and role models, and they contributed to a growing perception that these might be scientific disciplines where women could thrive. It takes time for women to reach this "critical mass" in a discipline such that their presence begins to contribute to developing a more hospitable climate (Rossiter, 1995).

Another reason is that our gendered assumptions about science and technology influence who participates: "The first term in the following pairs generally correlate with men, and the second with women: abstract/concrete, objectivity/subjectivity, logical/intuitive, mind/body, domination/submission" (Estrin, 1996, p. 44). This also informs the hierarchical ranking of academic fields accordingly. Fields such as engineering and computing are viewed as male domains because they are associated with hardness, machines, and abstraction. Fields such as biology and psychology are viewed as female domains because they are associated with softness, nature/people, and concreteness.

The ways in which this has impacted women's participation are evident in the historical access to some fields (over others) and the numbers of women in these fields today. The most feminized fields (those with the highest numbers of women) are biology and psychology—fields most closely associated with softness, nature, and people. The least feminized fields (those with the lowest numbers of women) are engineering and computer science—fields most closely associated with hardness and machines. In terms of percentages, the most feminized (biology and psychology) and least feminized (engineering and computer science) fields in 1970 remained so in 2000. In fact, psychology and biology accounted "for more than half of all doctorates awarded to women in all fields of science and engineering" (Rossiter, 2012, p. 95).

In 1970, women earned fewer than 2.5 percent of the doctorates in engineering and computer science, and in 2000 that number had only risen to 17 percent (Rossiter, 2012, p. 95). After the dot-com bust of 2000, numbers plummeted for men and women in computing, but women's numbers dropped off more dramatically (Rossier, 2012, p. 62). Engineering and computing have not reached the critical mass that makes it easier for women to enter and remain, in part due to gendered assumptions about the fields. Evidence that gendered assumptions still influence these fields today can be found in studies showing more women technology majors when the program is titled "computer science" and housed in Colleges of Arts and Sciences (perceived as softer) rather than when the program is titled "electrical engineering" and in Colleges of Engineering (perceived as harder; Cohoon & Aspray, 2006; Margolis & Fisher, 2002).

There is also an obvious hard/soft divide within computing itself—hardware (associated with the machine) and software (associated with people). There are more women software developers than hardware engineers; there are also more women in the people-focused (softer) information end of technology. Janet Abbate (2012) shares the history of women working on the first digital computers—ENIAC in the United States and Colossus in the United Kingdom—demonstrating how "assumptions about the gendered nature of technical skill, about women's place in the workforce, and about the nature of computing" constrained women's options and undervalued their contributions (p. 11). The hard/soft divide in computing today was evident among the participants in these two early computing projects, where men typically designed and built the hardware and women programmed the computers. Even though most of the women on both projects had math degrees from universities, there was an even further gender hierarchy on the Colossus project, where programming was defined in two parts: a mathematician (cryptographer) decided what operations the machine should perform, and an operator set up and ran the operations on the computer. All the cryptographers "were men, and all of the operators were women" (Abbate, 2012, p. 20).

CONCLUSION

Genderblindness is too deeply encoded in our knowledge tradition to be cured in one brief chapter. Becoming aware of the ways in which one individually makes assumptions based on gender, the ways in which technoscientific thought is encoded with gender (while claiming genderneutrality), and how that influences women's participation, and rediscovering the histories of women in science and technology, requires an ongoing commitment to the process of discovery. I hope these ideas have inspired you to engage in the process of becoming increasingly genderaware, because only then can we all work together to manifest our full human potential.

REFERENCES

Abbate, J. (2012). *Recoding gender: Women's changing participation in computing.* Cambridge, MA: MIT Press.

Barker, L. J., & Aspray, W. (2006). The state of research on girls and IT. In J. M. Cohoon & W. Aspray (Eds.), *Women and information technology: Research on underrepresentation* (pp. 3–54). Cambridge, MA: MIT Press.

Bix, A. S. (2000). Feminism where men predominate: The history of women's science and engineering education at MIT. *Women's Studies Quarterly, 28*(1/2), 24–45.

Bleier, R. (1991). *Feminist approaches to science.* New York, NY: Teachers College.

Bordo, S. (Spring 1986). The Cartesian masculinization of thought. *Signs, 11*(3), 439–456. doi: http://www.jstor.org/stable/3174004.

Cohoon, J. M., and Aspray, W. (Eds.). (2006). *Women and information technology: Research on underrepresentation.* Cambridge, MA: MIT Press.

Estrin, T. (1996). Women's studies and computer science: Their intersection. *IEEE Annals of the History of Computing, 18*(3), 43–46.

Feenberg, A. (2003). Technology in a global world. In R. Figueroa & S. Harding (Eds.), *Science and other cultures: Issues in philosophies of science and technology* (pp. 237–251). New York, NY: Routledge.

Figueroa, R., & Harding, S. (Eds.). (2003). *Science and other cultures: Issues in philosophies of science and technology.* New York, NY: Routledge.

Fine, C. (2010). *Delusions of gender: How our minds, society and neurosexism create difference.* New York, NY: Norton.

Frehill, L. M. (2004, April 1). The gendered construction of the engineering profession in the United States, 1983–1920. *Men and Masculinities, 6*(383). Retrieved from http://jmm.sagepub.com/content/6/4/383.

Harding, S. (1986). *The science question in feminism.* Ithaca, NY: Cornell University Press.

Harding, S. (1998). *Is science multicultural? Postcolonialisms, feminisms, and epistemologies.* Bloomington, IN: Indiana University Press.

Harding, S. (2006). *Science and social inequality: Feminist and postcolonial issues.* Urbana: University of Illinois Press.

Harding, S. (2008). *Sciences from below: Feminisms, postcolonialities, and modernities.* Durham, NC: Duke University Press.

Hartman, D. (Spring 2010). Insidious trauma caused by prenatal gender prejudice. *Journal of Heart Centered Therapies, 13*(1), 49. Retrieved from http://go.gale group.com.ezproxy.metrostate.edu.

Keller, E. F. (1983). *A feeling for the organism: The life and work of Barbara McClintock.* San Francisco, CA: Freeman.

Keller, E. F. (1985). *Reflections on gender and science.* New Haven, CT: Yale University Press.

Keller, E. F. (2010). *The mirage of a space between nature and nurture.* Durham, NC: Duke University Press.

Kirk, M. (2009). *Gender and information technology: Moving beyond access to co-create global partnership.* Hershey, PA: IGI Global.

Lerner, R. M. (2005). Foreword: Urie Bronfenbrenner: Career contributions of the consummate developmental scientist. In U. Bronfenbrenner (Ed.), *Making human beings human: Bioecological perspectives on human development* (pp. ix–xxvi). Thousand Oaks, CA: Sage.

Lohan, M., & Faulkner, W. (2004, April 1). Masculinities and technologies: Some introductory remarks. *Men and Masculinities, 6*(319). Retrieved from http://jmm .sagepub.com/content/6/4/319.

Margolis, J., & Fisher, A. (2002). *Unlocking the clubhouse: Women in computing.* Cambridge, MA: MIT Press.

Margolis, J., et al. (Eds.). (2010).*Stuck in the shallow end: Education, race, and computing.* Cambridge, MA: MIT Press.

Merchant, C. (1980). *The death of nature: Women, ecology and the scientific revolution.* San Francisco, CA: HarperSanFrancisco.

Millar, M. S. (1998). *Cracking the gender code: Who rules the wired world?* Toronto, Canada: Second Story.

Misa, T. J. (Ed.). (2010). *Gender codes: Why women are leaving computing.* Hoboken, NJ: Wiley.

Rossiter, M. W. (1982). *Women scientists in America: Struggles and strategies to 1940.* Baltimore, MD: Johns Hopkins University Press.

Rossiter, M. W. (1995). *Women scientists in America: Before affirmative action 1940–1972.* Baltimore, MD: Johns Hopkins University Press.

Rossiter, M. W. (2012). *Women scientists in America: Forging a new world since 1972.* Baltimore, MD: Johns Hopkins University Press.

Sadker, M., & Sadker, D. (1995). *Failing at fairness: How our schools cheat girls.* New York, NY: Touchstone.

Schiebinger, L. (1993). *Nature's body: Gender in the making of modern science.* Boston, MA: Beacon.

Schiebinger, L. (1999). *Has feminism changed science?* Cambridge, MA: Harvard University Press.

Stanley, A. (1995). *Mothers and daughters of invention: Notes for a revised history of technology.* New Brunswick, NJ: Rutgers University Press.

Toole, B.A. (1992). *Ada, the enchantress of numbers: A selection from the letters of Lord Byron's daughter and her description of the first computer.* Mill Valley, CA: Strawberry.

Wajcman, J. (2004). *Technofeminism.* Cambridge, UK: Polity.

Warren, W. (1999). *Black women scientists in the United States.* Bloomington: Indiana University Press.

Williams, S.W. (1999, January 1). History of black women in the mathematical sciences. Retrieved from http://www.math.buffalo.edu/mad/wohist.html.

NOTES

1. In accordance with science and technology studies scholars, I use the term *technoscience* rather than *science* or *technology* as a marker of the ways in which these knowledge traditions are socially situated.

2. Thanks to Cordelia Fine's (2010) *Delusions of Gender: How Our Minds, Society and Neurosexism Create Difference* for this title.

3. This is a big project for one chapter. For a more in-depth explanation of some of these issues, see my (2009) book *Gender and Information Technology: Moving Beyond Access to Co-Create Global Partnership.* I'm currently working on a new book.

SUSTAINABILITY

CHAPTER 6

ECO-TECHNOLOGICAL
LITERACY FOR RESILIENCY

Leo Elshof

In terms of education, Rifkin suggests that "our ideas about education invariably flow from our perception of reality and our conception of nature—especially our assumptions about human nature and the meaning of the human journey. Those assumptions become institutionalized in our education process. What we really teach, at any given time, is the consciousness of an era" (2011, p. 234).

This has significant implications for the sustainability crisis that envelops the planet. It is primarily a symptom of human ecological dysfunction: our propensity for cognitive dissonance and collective denial concerning the enormity of our impact on the ecosystems that sustain us. The environmental crisis reflects our collective dysfunctional consciousness, a consciousness that perpetuates a profound misunderstanding of who and what we are—that is, only one of many interdependent animal species living on a very finite planet that is rapidly becoming "human full" (Rees, 2010). Unfortunately, the consciousness of a bygone era still clings to our conceptions of technological literacy to the extent that ecological transparency and mindful attention to issues of equity and environmental justice used to inform the design, manufacture, and use of technologies remains marginalized, underemphasized, or simply neglected altogether. Technological education is often the only opportunity in formal education for young people to enter into a sustained and critically reflective relationship with physical materials, tools, processes of design and manufacture, and the human ecology dimensions of design and consumption. This chapter will examine the nature of eco-technological literacy and the urgency in advancing it as an integral component of twenty-first-century literacy.

Progress and Its Long-Term Cost

Technological development has played an enormous role in expanding human progress. Over the past 40 years in 135 countries representing 92 percent of the world's population, average life expectancy rose from 59 to 70 years, primary school enrolment grew from 55 to 70 percent, and per capita income doubled to more than US$10,000 (United Nations Development Programme [UNDP], 2011). We are just beginning to understand the very significant environmental costs associated with achieving these very laudable goals. The UN secretary general is blunt:

> For most of the last century, economic growth was fuelled by what seemed to be a certain truth: the abundance of natural resources. We mined our way to growth. We burned our way to prosperity. We believed in consumption without consequences. Those days are gone. In the twenty-first century, supplies are running short and the global thermostat is running high. Climate change is also showing us that the old model is more than obsolete. It has rendered it extremely dangerous. Over time, that model is a recipe for national disaster. It is a global suicide pact. (Ki-Moon, 2011)

In short, we are simply living beyond the biocapacity of the planet, and our deficit spending is growing larger. The "Living Planet Index" measures changes in the health of the planet's ecosystems by tracking post-1970 trends of more than 9,000 populations of 2,688 species of birds, mammals, amphibians, reptiles, and fish. The index reveals a decline in health of approximately 30 percent from 1970 to 2008 (World Wildlife Fund [WWF], 2012). Analysis of our global ecological footprint, a measure of our aggregate impact on the planet, reveals that the earth is taking 1.5 years to fully regenerate the renewable resources that people are using in one year, so instead of living off the interest, we are actually eating into the natural capital of the planet (WWF, 2012).

Attempting to raise all the planet's citizens to the enormous material consumption levels enjoyed in North America would require the biocapacity of four planets: a physical impossibility (Rees, 2010). Despite this, all major economies are firmly committed to this trajectory. Although major scientific studies have rediscovered truths long held by many indigenous peoples—namely, that poverty alleviation, protecting human health, and ensuring long-term prosperity all critically depend on maintaining the flow of benefits from healthy ecosystems (Sukhdev, 2010)—we continue to degrade them in the name of short-term economic progress. Global carbon dioxide emissions in 2012 were the second highest on record (National Oceanic and Atmospheric Administration [NOAA], 2013), and hopes to stabilize the global average temperature increase to no more than 2°C, the threshold level for "dangerous" climate change, are rapidly receding (Kirby, 2013). In fact, greenhouse emissions are rising faster than the worst-case scenario projections laid out by the Intergovernmental Panel on Climate Change.

Whether we are prepared or not, human beings—the primary change agents on the planet, architects of their own "Anthropocene"—are being drawn into becoming "active stewards of our own life support system." As Steffen and Persson explain, "The Anthropocene is a reminder that the Holocene, during which complex human societies have developed, has been a stable, accommodating environment and is the only state of the Earth System that we know for sure can support contemporary society. The need to achieve effective planetary stewardship is urgent. As we go further into the Anthropocene, we risk driving the Earth System onto a trajectory toward more hostile states from which we cannot easily return" (2011, p. 739).

As Ehrlich, Kareiva, and Daily (2012, p.68) conclude, "In biophysical terms, humanity has never been moving faster nor further from sustainability than it is now," raising questions as to whether humanity can steer clear of major societal collapse in the near to long-term future. If this "litany" sounds familiar, it is because it is. More than twenty years ago in 1992, the United Nations Rio Conference summary document, Agenda 21, concluded, "The major cause of the continued deterioration of the global environment is the unsustainable pattern of consumption and production" (UN Agenda 21, Chapter 4, 1992).

The story remains little changed today. The UN Human Development Index (HDI) is a measure of well-being. The average person in a very high HDI country produces more than four times the carbon dioxide emissions and about twice the methane and nitrous oxide emissions of a person in a low, medium, or high HDI country. High HDI citizens produce about thirty times the carbon dioxide emissions of a person living in a low HDI country (United Nations Development Programme [UNDP], 2011). Addressing this inequity involves creating more "ecological space" for the poorest inhabitants to enjoy more material growth, while simultaneously shrinking the ecological footprint of the world's richest countries.

But it is not only ecological systems that are threatened by our consumption behaviors; a number of the world's rare earth minerals are being depleted at an alarming rate. Minerals such as indium, tantalum, and platinum, which are crucial for the manufacture of things like flat-screen LCD monitors, catalytic converters, photovoltaic panels, and fuel cells, are being rapidly depleted (Cohen, 2007). Across the planet, governments and transnational corporations are aggressively pursuing whatever mineral and agricultural resources they can in remote and ecologically sensitive regions. As Klare explains, these groups recognize that "existing reserves are being depleted at a terrifying pace and will be largely exhausted in the not-too distant future. The only way for countries to ensure an adequate future supply of these materials, and thereby keep their economies humming, is to acquire new, undeveloped reservoirs in those few locations that have not already been completely drained. This has produced a global drive to find and exploit the world's final resource reserves—a race for what's left" (2012, p.11).

The trends described here help set the background context for considering the idea of eco-technological literacy. I now turn to briefly examine the social psychology that lies behind the "race for what's left."

Toward an Understanding
of Our Human Predicament

Our consumption-driven economies are founded on economic principles that treat the planet's ecosystems as a bottomless resource pool and as a convenient place to dump the byproducts of human technological activity that we have not yet imagined a use for. The logic of neoliberal production systems and the material culture that it has helped create are nothing short of a recipe for planetary ecological disaster.

Neoliberal capitalism is, at its core, driven by what White terms the "barbaric heart," which believes that prosperity is dependent on violence and that "if you can prosper from violence, then you should go ahead and be violent" (2009, p.9). Violence in this context is not limited to physical violence against other humans in terms of overconsuming our fair share of planetary resources but also includes the systemic structures of oppression and violence our economic, consumption, and technological systems wreak on the long-term well-being of the earth's ecosystems. In effect, we commit violence against the future well-being of generations to come.

White contends that we often frame the attributes of the barbaric heart as a form of virtue, "especially if you think that winning, surviving, triumphing and accumulating great wealth are virtues, just as athletes, Darwinians, military commanders, and capitalists do" (2009, p.9). The barbaric heart is prereflective and, as White warns, "is no better at questioning itself about the meaning of its actions than capitalism is at asking why the growth of the Gross Domestic Product is good. Capitalism does not ask, what's the economy for? It merely asks it to grow" (2009, p. 8).

The neoliberal consumption dynamic informed by the barbaric heart is at the core of the sustainability crisis. The logic of the barbaric heart is reproduced in education contexts that take as a given its so-called virtues, which include an overemphasis on hypercompetition, narrow neoliberal notions of economic efficiency, and the development of instrumental skill. A focus on these attributes should not come at the cost of neglecting critical and sustained reflection on the dynamics of consumption and production from multiple perspectives, including environmental justice and equity. It is important for technological education to help students develop a sense of critical discernment and judgment concerning the "economic imperative" arguments made for the continued or expanded manufacture and use of highly polluting and/or material- and energy-intensive technologies.

As Rifkin argues, Enlightenment thinking at the dawn of the age of reason posited human beings' essential nature as "rational, detached, autonomous, acquisitive and utilitarian," and it was thought that "individual salvation lies in unlimited material progress here on Earth" (2011, p. 3). Neoliberal economic thinking is firmly rooted in this understanding of the individual, and the technological systems and stories that disregard the environmental impacts from high levels of material and energy consumption are based on serving the needs and wants of this same old Enlightenment notion of human nature. Underpinning neoliberal economics is a mental model that posits human beings as being separate or detached from the health of the biosphere and the animal world and their societies and economies as distinct, even independent of, the planet's natural systems. This is the central flawed mental model and the source of irrationality deeply implicated in the sustainability crisis that envelops the planet. As Rifkin warns, "If human nature is as many of the Enlightenment philosophers claimed, then we are likely doomed. It is impossible to imagine how we might create a sustainable global economy and restore the biosphere to health if each and every one of us is, at the core of our biology, an autonomous agent and a self-centered and materialistic being" (2011, p. 2).

The millions of disposable "products" of Enlightenment thinking described by Rifkin literally fill the "big-box" retail stores and landfills of North America. These ephemeral technological products are designed and built from the familiar antiquated linear model of manufacture, consume, and dispose. Hundreds of thousands of products are sold with little or no transparency regarding the nature or sources of the raw materials or energy consumed in their production and manufacture, or for that matter any longer-term life cycle considerations or repairability/reuse after the product has served its useful life. Worldwide, tens of millions of different manufactured products (Jackson, 2010) are produced with material, energy, labor, and pollution histories that are largely unknown to the consumer at the end of the supply chain.

With no social history and no connection to geographical place, we have no opportunity and feel no need to develop a long-term "relationship" with a product that extends beyond our initial novelty fascination with it or after it has served its often short-lived purpose. We quickly move on to the next indispensable "new" thing, cajoled by a billion-dollar marketing system that ensures that our sense of satisfaction with a product is as short lived as the product itself. Is it any surprise that young people who have never been encouraged to develop a sustained critical engagement or relationship with their own "story of stuff" (Leonard, 2010) develop a cavalier, disposable attitude toward their world of material abundance? The neglected "back story" that needs to come to the forefront includes the nature, source, and longevity of materials; the quantity and nature of the energy that is embodied in materials; the waste and byproducts that are released throughout the lifecycle; the skill levels of the workers; and the working conditions of the people involved in the supply chains that stretch around the planet. As Saul reminds us, "Our challenge is to learn how to recognize what we have trained ourselves not to see. We must remove the imaginative and historical veils that we have used to obscure this reality" (2005, p. 35).

Eco-technological literacy has, for the most part, been a relatively minor component of technological education and a latecomer at that. While official curricula have improved over the last decade, insofar as many now mention that students will "examine" or "consider" the environmental "impacts" of technological decision making, most remain woefully inadequate in terms of specifically identifying how this analysis will be carried out using quantifiable and qualitative tools. Although environmental "concern"or other rhetorical "sustainability"placeholders within the curriculum may be well meaning, without specifics and focused learning activities that confront the quantifiable and holistic facts surrounding pollution, consumption, and the profligate use of materials and energy, they reflect a degree of insincerity and superficiality: "The truth is that without significant precautions, education can equip people merely to be more effective vandals of the earth" (Orr,1994, p. 5).

Although eco-technological practices have advanced in many industries over the last twenty years, there remain very powerful groups that oppose ecological transparency and a "polluters-pay" accountability. For example, despite the extremely serious scenario of climate change, neoliberal corporate groups like the National Association of Manufacturers (NAM, 2013) and the American Chamber of Commerce, the largest industry and business and lobbying groups in North America, along with similar groups in Canada, oppose initiatives like a "carbon tax"or a "cap-and-trade"system in order to protect their privilege, power, and profit. Billionaires like the petrochemical tycoon Koch brothers (Hartmann & Sacks, 2012) and other conservative groups in the United States attack pro–clean energy policies (Pernick, Wilder, & Winnie,2013) in the name of their primary funders: the oil, coal, and gas industries. It is important to emphasize that these organizations effectively promote a form of "moral blindness" (Bauman & Donskis, 2013), an indifference to an environmental and social justice considerations of their business practices in the name of profit. They would rather pass on these costs to other citizens and future generations.

In Canada, oil industry consortia worked for many years to finally defeat the Kyoto Protocol on climate change, and they produce public relations educational materials that underemphasize the catastrophic impact of climate change on Canada's ecosystems and peoples. In the United States, corporate consortia funded by billionaires such as the "American Legislative Exchange Council"produce mandates for lawmakers that advocate the "balanced teaching" of climate science in public school classrooms. As Chomsky explains, "'Balanced teaching' is a code phrase that refers to teaching climate-change denial, to 'balance' mainstream climate science. It is analogous to the 'balanced teaching' advocated by creationists to enable the teaching of 'creation science' in public schools" (2013). The purpose here is simple: to spread disinformation concerning the real impacts of our hydrocarbon-intensive technologies and lifestyles and to discredit those working to help us avoid a climate catastrophe.

It is important that young people understand that those who don't want the environmental wake of their way of doing business to be publicized or penalized wield an enormous influence on media and political decision making. There has never been a more pressing need to educate young people about the systemic irrationality built into current conceptions of neoliberal production-consumption economics. Jackson explains how "fast consumption" is the inevitable partner of "fast production":

If we don't consume, then who is going to produce? And without production who will employ us? Without jobs how will we maintain our ability to go on and on consuming? Indeed the underlying dynamic here is not just about continuing to consume, but about consuming more and more. The stability of the economy itself in the 'advanced' consumer societies calls on us not simply to maintain our productive capacity but to pursue a strategy of continuing, exponential growth. The dynamic that feeds this strategy relies on the relentless production of novelty by firms and the relentless consumption of novelty by households. The inevitability with which this leads to a throwaway culture is patent. (2010. p. xv)

Fortunately, a number of emerging interdisciplinary methodological tools, concepts, and approaches are edging into design and hopefully technological education. These tools address various facets of the sustainability conundrum and as a whole encourage us to think more holistically about technology. They include the following:

- Design for the Environment (DfE)
- Life-cycle analysis (LCA) and life-cycle management
- Industrial ecology and material flow analysis
- Carbon and ecological footprinting
- Design for disassembly, recycling, and remanufacture
- Eco-efficiency and biomimicry
- ISO 14001 and environmental management
- The Natural Step and "triple bottom line" accounting

The next generation of technologists needs to create a "discontinuous leap" (Ehrenfeld, 2008) from the existing high-consumption product forms of technoculture. Their capacity to develop creative democratic and participatory strategies to bypass the old in order to create a new eco-technological reality will largely determine their quality of life going forward. The task for educators involves mobilizing the green, eco-technological imagination of young people as an antidote to "doom and gloom" inevitability. This involves developing technological literacies that provide insight into how technologies can sustain genuine human flourishing along with authentic forms of technological design that extend beyond the instrumental commodified forms of the "better living through more products"mantra that corporate media intensively sells. Modernity, suggests Ehrenfeld, has "dimmed" three critical domains that are essential for human flourishing and true sustainability:

- Our sense of ourselves as human beings: the human domain.
- Our sense of our place in the [natural] world: the natural domain.
- Our sense of doing the right thing: the ethical domain. (Ehrenfeld, 2008, p. 58)

Eco-technological education can be a reconstructive journey back to a deeper consideration of these facets of sustainability. All involve developing learning opportunities for students to develop a more mindful connection with our technological creations. Design mindfulness involves a determination to

- think about the consequences of design actions before we take them and pay close attention to the natural, industrial, and cultural systems that are the context of our design actions;
- consider material and energy flows in all the systems we design;
- give priority to human agency and not treat humans as a "factor" in some bigger picture;
- deliver value to people—not deliver people to systems;
- treat "content" as something we do, not something we are sold;
- treat place, time, and cultural difference as positive values, not as obstacles; focus on services, not on things, and refrain from flooding the world with pointless devices. (Thackara, 2005, p. 8)

It is worth reminding ourselves that sustainability is a normative concept; it remains an aspirational goal that needs to be defined in terms of a local and bioregional context. Sustainability is "an emergent property of a complex system; we can observe it only if all the relationships on which it depends are functioning correctly" (Ehrenfeld, 2008, p. 65). This entails connecting students to the stories of individuals and organizations who are working under different precepts and visions than those being promoted by transnational corporations and retailers. There are many exemplary stories of young people changing the world for the better through technological and social innovation. For example, the stories of creative young people from 5 continents and 35 countries working toward the UN Millennium Development Goals are documented in the design science/global solutions lab project (Gabel, 2010).

Goleman explains "radical transparency"in terms of what it means for consumers equipped with the full disclosure of information related to the full environmental and social costs of products: "Radical transparency converts the chains that link every product and its multiple impacts—carbon footprints, chemicals of concern, treatment of workers, and the like—into systematic forces that count in sales. Radical transparency leverages a coming generation of tech applications, where software manipulates massive collections of data and displays them as a simple readout for making decisions. Once we know the true impacts of our shopping choices, we can use that information to accelerate incremental changes for the better" (2009, p. 10). While the radical transparency described by Goleman is as yet unrealized, and despite the powerful forces advocating for the status quo of "ecological blindness," radical transparency signifies a key aspirational goal for eco-technological literacy. Young people need to begin working with the best available knowledge, including ecological and carbon footprints, and full life-cycle analyses in all phases of technological education.

"Design activism"attempts to create a new counternarrative to the neoliberal modernist product culture. It incorporates ecological and social awareness into all facets of technological design. As Fuad-Luke explains, "Design activism is 'design thinking, imagination and practice applied knowingly or unknowingly to create a counter-narrative aimed at generating and balancing positive social, institutional, environmental and/or economic change'" (2009, p. 20).

New design paradigms are embracing emerging modes of participatory culture, with customers becoming cocreators and codesigners. Codesign is striving to be more "democratic, open and porous" as it engages and gives a voice to the people who will use a given technology in the design process (Fuad-Luke, 2009, p. 147). Today it is crucial that students understand not only global material flow processes but also the local contextualized dimensions of science-technology-society-environment relationships. This entails that students develop a place-based understanding of how economic, technoscientific, and social policies contribute to or detract from environmental sustainability and local quality of life.

The world of the near future will be characterized by increasing levels of social and technological complexity (trends difficult enough to cope with) but also by a growing world population and increasingly compromised ecosystems. These trends will combine to make large-scale and often unintended and uncontrollable disturbances the new norm. Educational systems will need to prepare students who can effectively deal with accelerating levels of uncertainty and change. In short, societies will need more resilient learners who can design more resilient eco-technological systems. In its simplest sense, *resilience* is "the capacity of a system to absorb disturbance and still retain its basic function and structure" (Walker & Salt, 2006, p. xi). The key questions young people need to grapple with are the following: How can we imagine our technologies contributing to the development of a more resilient world? How do our technologies advance us toward, or move us away from, resiliency? The basic characteristics of a "resilient world" include the following:

1. Diversity—A resilient world would promote and sustain diversity in all forms (biological, landscape, social, and economic).
2. Ecological variability—A resilient world would embrace work with ecological variability (rather than attempting to control and reduce it).
3. Modularity—A resilient world would consist of modular components.
4. Acknowledging slow variables—A resilient world would have a policy focus on "slow" controlling variables associated with thresholds.
5. Tight feedbacks—A resilient world would possess tight feedbacks (but not too tight).
6. Social capital—A resilient world would promote trust, well-developed social networks, and leadership (adaptability).
7. Innovation—A resilient world would place an emphasis on learning, experimentation, locally developed rules, and embracing change.
8. Overlap in governance—A resilient world would have institutions that include "redundancy" in their governance structures and a mix of common and private property with overlapping access rights.

9. Ecosystem services—A resilient world would include all the unpriced eco-system services in development proposals and assessments (Walker & Salt, 2006, p. 146).

Technological education focused on building resiliency also understands that "all design is social"and the important distinction between "design for the market,"where the primary focus is creating products to sell, and "social design,"a process of satisfying real human needs while improving well-being and livelihood, in effect using design to change "existing situations into pre-ferred ones" (Fuad-Luke, 2009, p. 152): "Aspiring design activists have to be prepared to take on multiple roles as nonaligned social brokers and catalysts, facilitators, authors, co-creators, co-designers and 'happeners' (i.e. making things actually happen)" (Fuad-Luke, 2009, p. 189).

Global movements such as the "Natural Step" and the "Transition Initia-tive" (TransitionNetwork.org, 2013) are excellent examples of how com-munities are starting small, local grassroots projects to respond to the global challenges of climate change, economic hardship, and dwindling supplies of cheap energy. They bring together local expertise to work cooperatively to transition communities away from high carbon energy use, material con-sumption, and ecological impact toward a more sustainable and healthy alter-native. As the designer Thackara emphasizes, "The creation of interesting social alternatives has to be as exciting and engaging as the buzz of new technology used to be. A culture of community and connectivity has to be fun and challenging, as well as responsible. An aesthetics of service and flow should inspire us, not just satisfy us" (2005, p. 8).The challenge for educa-tors will be to design community learning experiences with eco-technologies that young people find both inspirational and meaningful.

CONCLUSION

Advancing eco-technological literacy is not about deprivation or a return to old craft ways of designing and working with technology; rather, it is about innovation, collaboration, and connectivity. It involves the conscious choice to realign our technological skills and innovation to support as best we can the proper functioning of our planetary ecosystems, because it is in our best interests to do so. This will perhaps be the ultimate intelligence test for humankind. As Sawhill states, "In the end, our society will be defined not only by what we create, but by what we refuse to destroy" (2006).

REFERENCES

Bauman, Z., & Donskis, L. (2013). *Moral blindness: The loss of sensitivity in liquid modernity*. New York, NY: Polity-Wiley.

Chomsky. N. (2013). Can civilization survive capitalism? Retrieved March 5, 2014, from http://www.alternet.org/noam-chomsky-can-civilization-survive-capitalism.

Cohen, D. (2007, May 23). Earth's natural wealth: An audit. *New Scientist.*

Ehrenfeld, J. R. (2008). *Sustainability by design.* New Haven, CT: Yale University Press.

Ehrlich, P. R., Kareiva, P. M., & Daily, G. C. (2012). Securing natural capital and expanding equity to rescale civilization. *Nature, 486*(7401), 68–73.

Fuad-Luke, A. (2009). *Design activism: Beautiful strangeness for a sustainable world.* London, UK: Earthscan.

Gabel, M. (2010). *Designing a world that works for all: How the youth of the world are creating real-world solutions for the UN Millennium Development Goals and beyond.* New York, NY: CreateSpace Independent Publishing Platform.

Goleman, D. (2009). *Ecological intelligence.* New York, NY: Broadway Books.

Hartmann, T., & Sacks, S. (2012). The case against billionaires. Retrieved January 3, 2012, from http://truth-out.org/opinion/item/13698-the-case-against-billiona.

Jackson, T. (2010). Foreword. In T. Cooper (Ed.), *Longer lasting products* (pp. xv–vii). Farnham, Surrey, UK: Gower.

Ki-Moon, B. (2011). Twentieth-century model: "A global suicide pact." Retrieved March 2014, from http://www.un.org/News/Press/docs/2011/sgsm13372.

Kirby, A. (2013, March 6). Coal triggers carbon level rise. Climate News Network. Retrieved March 14, 2014, from http://www.climatenewsnetwork.net/2013/03/coal-triggers-carbon-level-rise.

Klare, M. T. (2012). *The race for what's left: The global scramble for the world's last resources.* New York, NY: Metropolitan Books.

Leonard, A. (2010). *The story of stuff.* New York, NY: Simon and Schuster.

National Association of Manufacturers (NAM). (2013). *Economic outcomes of a U.S. carbon tax.* Washington, DC: Washington National Association of Manufacturers, NERA Economic Consulting.

National Oceanic and Atmospheric Administration (NOAA). (2013). Trends in atmospheric carbon dioxide. Retrieved March 7, 2013, from http://www.esrl.noaa.gov/gmd/ccgg/trends.

Orr, D. (1994). *Earth in mind: On education, environment, and the human prospect.* Washington, DC: Island Press.

Pernick, R., Wilder, C., & Winnie, T. (2013). Clean energy trends 2013. Clean Edge. Retrieved March 12, 2014, from http://www.cleanedge.com.

Rees, W. (2010). What's blocking sustainability? Human nature, cognition, and denial. *Sustainability: Science, Practice, & Policy, 6*(2), 13–25.

Rifkin, J. (2011). *The third industrial revolution.* New York, NY: Palgrave Macmillan.

Saul, J. R. (2005). *The collapse of globalism and the reinvention of the world.* Toronto, Canada: Viking Canada.

Sawhill, J. (2006). *Congressional Record, 152*(7), 8716.

Steffen, W., & Persson, A. (2011). The Anthropocene: From global change to planetary stewardship. *AMBIO, 40,* 739–761.

Sukhdev, P. (2010). Putting a value on nature could set scene for true green economy. *Guardian.* Retrieved March 2014, from http://www.theguardian.com/commentisfree/cif-green/2010/feb/10/pavan-sukhdev-natures-economic-model.

Thackara, J. (2005). *In the bubble: Designing in a complex world.* Cambridge, MA: MIT Press.

TransitionNetwork.org. (2013). About Transition Network. Retrieved March 14, 2014, from http://www.transitionnetwork.org.

United Nations. (1992, June 3–14). Agenda 21: Report of the United Nations conference on environment and development. Retrieved from http://www.un.org/documents/ga/conf151/aconf15126-4.htm.

United Nations Development Programme (UNDP). (2011). Towards a green economy: Pathways to sustainable development and poverty eradication. http://www.unep.org/greeneconomy.

Walker, B., & Salt, D. (2006). *Resilience thinking*. Washington, DC: Island Press.

White, C. (2009). *The barbaric heart*. Sausilito, CA: Polipoint Press.

Worldwide Wildlife Fund (WWF). (2012). *Living planet report 2012*. Gland, Switzerland: WWF International, Zoological Society of London, Global Footprint Network.

Technological Literacy in China

CHAPTER 7

A CHINESE PERSPECTIVE ON
TECHNOLOGICAL LITERACY

Nan Wang

BACKGROUND: LITERACY IN CHINA

Literacy involves formal education. Learning to read and write in Chinese is inherently more time consuming than it is in nonideographic languages, and some degree of formal education has been more continuous in China than in any other society in the world. Although the *Analects* describe conversations between the sage Confucius and his students in a time roughly parallel to that presented in Plato's Socratic dialogues, Plato's Academy was closed in 529CE and never reopened. By contrast, Confucian educational traditions, although often interrupted, have repeatedly been revived and continue into the present.

Additionally, prior to the modern period, Chinese philosophy was the foundation for all formal education. Even the earliest grades stressed the reading and memorizing of passages from the Confucian classics. Philosophy could serve this function in primary education because, in the Confucian tradition, learning provides guidance for dealing with the problems of human life. Confucius expressed such a view clearly at the beginning of the basic Confucian text: "To learn and then have occasion to practice what you have learned—is this not satisfying?" (Kongzi, 2003, p.1). The goal of Confucian philosophy is practical, not theoretical, knowledge.

The nonelitist and practical character of Chinese philosophy can be elaborated with three further points. First, it should be noted that the Western word *philosophy* is quite recent in Chinese; it was initially rendered into Chinese as *zhexue* in 1873 by the Japanese scholar Xi Zhou (1829–1897), who studied in the Netherlands. Recall that, from 1637 to 1854, the only Western country in regular contact with Japan was the Netherlands and that in the late 1800s Japan was also of considerable influence in China, especially because of the positive Japanese example of learning from the West. According to *Shuowen jiezi* (Analytical Dictionary of Characters), the first word book giving a systematic analysis of grapheme and word origins in Chinese, the initial character *zhe* means "knowledge" or "capacity to acquire knowledge," with an extended meaning of "wisdom"; the second character, *xue*, means "learning." The Chinese term for *philosophy* thus means "learning to become a wise and knowledgeable person" (Cua, 2003, p. 499). Xi Zhou came up with his proposal to translate "philosophy" into Chinese only after a long period of reflection on the best way to capture the meaning of Western philosophy in the Chinese language. Before Xi Zhou's word coinage, the abstract notion of philosophy was always embedded in more specific neo-Confucianist terms created by Cheng Hao (1033–1107), Cheng Yi (1032–1085), and Zhu Xi (1130–1200), such as *qiongli xue* (inquiry learning into the universe), *xingli xue* (theory of human nature), or *li xue* (learning of principle).

In a note to explain why he decided to create *zhexue* to replace specific terms with a more general one, Xi Zhou wrote:

> The original English word for zhexue is philosophy, and the French word is philosophie. Both derive from the Greek word philosophos, which means the person who loves (philo) wisdom (sophos). The functional implication in the Chinese language is the so called "scholar who follows the example of the wise person," according to a proposal by Zhou Dunyi in the Song dynasty. Later generations specifically identified philosophy [in general] with neo-Confucianism [a specific philosophy] and even literally translated the former as the doctrine of neo-Confucianism. In many instances, it is better to translate philosophy as zhexue in order to distinguish it from Confucianism in East Asia. (Sun, 2010, p.125)

This analysis by Xi Zhou requires commentary. Zhou Dunyi (1017–1073) was a Chinese neo-Confucian philosopher who, in his *Tongshu* (All-Embracing Book), distinguished three types of educated persons: the sage, the wise person, and the scholar. The sage is the most educated and acts in accordance with the principles of heaven (i.e., of all reality); the wise person strives to be but has not yet become a sage and is thus of a lower rank; and the scholar strives to be but has not yet become wise and is of still lower rank. In Confucianism, the ultimate goal of education is nothing less than sagehood; however, this cannot be achieved overnight but only through a step-by-step transition.

Second, it can also be noted that even when acting in accordance with the principles of heaven, the sage is not separated from human affairs. In the words of Feng Youlan (1895–1990), a Chinese philosopher who made special contributions to the revitalization of Chinese philosophy in the twentieth century, the sage stands out not in terms of behavior but in terms of orientation. According to Feng, "The sage does nothing more than most people do, but, having high understanding, what he does has a different significance to him. In other words, he does what he does in a state of enlightenment, while other people do what they do in a state of ignorance" (2007, p.560).

Finally, in Chinese philosophy, ethics dominates over epistemology: practice over theory. In the words of a knowledgeable and influential American interpreter of Chinese philosophy,

> The Platonists were more concerned with knowing in order to understand, while the Confucians were more concerned with knowing in order to behave properly toward other men.
>
> In China, truth and falsity in the Greek sense have rarely been important considerations in a philosopher's acceptance of a given belief or proposition; these are Western concerns. The consideration important to the Chinese is the behavioral implications of the belief or proposition in question. What effect does adherence to the belief have on people? What implications for social action can be drawn from the statement? (Munro, 2001, pp. 54–55)

Confucius makes the same point in the following passage: "Every day I examine myself on three counts: in my dealings with others, have I in any way failed to be dutiful? In my interactions with friends and associates, have I in any way failed to be trustworthy? Finally, have I in any way failed to repeatedly put into practice what I teach?" (Kongzi, 2003, p. 1).

This strong practical orientation naturally supports an appreciation of human practice and experience. Technics—the craft making and use of artifacts, which has become systematized into technology in the modern period—is an indispensable part of daily life. As such, technics naturally becomes very early an object of Chinese philosophy. This is reflected in the various words used to discuss making and using.

In modern Chinese, English words such as "art," "skill," "technique," and "technology" can all be translated as *jishu*. Unlike the situation with philosophy as *zhexue*, the creator of the word *jishu* is unknown. But *jishu* is also a word that deserves special comment. In ancient Chinese, *ji* and *shu* were always used separately. According to *Shuowen jiezi*, *ji* means "ingeniousness and skillfulness of craftsman," with an extended meaning of "(exclusive) talent and the ability of craftsmen in general," although it sometimes refers to "certain special arts," such as singing and dancing. *Ji* can be acquired only by intuition and understanding and can only be perfected through practice. The original meaning of *shu* is "the ways or roads in the town," with an extended meaning of "skill, method, procedure." *Shu* refers not only to the skill, method, and process in physically making and using but also to mental action, political trickery, martial arts, art, mathematical calculation, necromancy, Daoist magic, and more. In this sense, Chinese knowledge is based on *shu*, which means that it pays more attention to the configuration of methods and procedures in order to memorize and be able to use them flexibly in practice (Liu, 1996, p. 688).

There is still a third word that, because of its close association with "technology," deserves mention: *gongcheng*, or "engineering." According to *Shuowen jiezi*, the original meaning of *gong* is literally "(artisan's) skillful work on adorning something," but some scholars, such as the Chinese linguist Yang Shuda, argue that *gong* originally referred to "a kind of instrument, such as bevel gauge" (Yang, 1954, p.58) and that *cheng* was "a measurement unit of length," with an extended meaning of "a general name of measurements of all kinds." As historians Joseph Needham and Wang Ling have concluded, "From the earliest times the word gong implied work of an artisanal character, technical as opposed to agricultural. This is perpetuated in the modern term for engineering, gongcheng, the second of the two characters having originally meant measurement, dimension, quantity, rule, examination, reckoning, etc." (Needham, 1963, p. 9).

According to Yang Shengbiao and Xu Kang, the association of the two words appeared at the latest in *Xin tang shu* (New History of the Tang Dynasty; 1060). They argued that *gongcheng* usually referred to the standard or evaluation of the progress of skilled work, especially the manual work schedule (Yang and Xu, 2002, p. 38). The fact that traditional engineering was chiefly civil engineering (e.g., the construction of imperial palaces and official offices, temples, canals, city walls, bridges) means that, as Needham and Wang also observed, "the associations of the Chinese terms for engineers and artisans seem always to have been more civilian and less military than [similar terms] in the West" (Needham, 1963, p. 10). *Gongcheng* became the translation for the English "engineering" in China during the self-strengthening movement (1861–1895) by British missionary and scholar John Fryer (1839–1928), who, among the foreigners in China, translated the greatest number of western books into Chinese (Wang and Xu, 2002, p. 39).

Generally speaking, from ancient times, philosophy in China was always associated with more clearly practical interests than it was in the West. Moreover, technology and engineering paid more attention to the artisanal character related to civil facilities than they did in Europe, where technical activity was so often associated with the military. These various points all support the idea of a unique philosophical perspective on technology and technological literacy in China.

PHILOSOPHY AND TECHNICS IN PREMODERN CHINA

To understand the philosophy of technics in ancient China, it is necessary to have some appreciation of the basis of Chinese culture. Geographically, China is a land-based country; seafaring and maritime trades are not nearly as important to China as they were, for instance, to Greece and Rome. Chinese life is based in agriculture. Even as late as the beginning of the twenty-first century, almost 50 percent of the Chinese population remained engaged in farming. It is no wonder that agricultural technics were originally the subject of Chinese philosophical reflection.

The earliest Chinese text to deal with technics is generally considered to be *Kaogongji* (Records of Examination of Artisans), an official book of the Qi state, one of a dozen small states during the Spring and Autumn

Period (770–476 BCE). It was composed by an unknown author and describes manufacturing processes and specifications related to carpenters, metalsmiths, leatherworkers, dyers, jewelers, and potters. But it elaborates especially on the methods of use of various agricultural and handicraft tools, which it classifies into diverse categories. Regarding the making of artifacts, the *Kaogongji* says, "The heaven has seasonal and climate change, the land has geographical differences, materials have various properties, and artisans have different types of creativity and skill. Gathering these four can produce good products" (Wen, 1993, p. 5).

The *Kaogongji* offers an interesting Chinese contrast to Aristotle's four causes (material, formal, final, and efficient). Whereas Aristotle's four causes are derived from reflection on human fabrication and then projected into the world of nature, the *Kaogongji* limits itself to aspects of the world that the human worker needs to take into consideration. There is no attempt to apply the Chinese analysis to reality as a whole or to give it metaphysical significance. This again reflects a distinctive this-worldly commitment in Chinese experience and culture.

Another contrast with traditional views of technics in the West can be found in a passage from the classic Daoist text attributed to Zhuangzi (who flourished during the Warring States Period, 476–221 BCE). In general, Daoists are critical of urban life and thus, by implication, of technics and especially artifacts. But one persistent tendency in the West is to fail to appreciate the inherent goodness of technics, independent of the skill of the artisan or the products created—an appreciation that can be found even in the sometimes radical antitechnical culture of philosophical Daoism.

Daoists all agreed that the dao is the supreme principle of the world. But Zhuangzi converted it from just an ontological principle to something also manifested in daily life (Chen, 1999, pp. 77–78), because he believed that the dao, as the origin of all things, must be reflected in ordinary human activities. Zhuangzi called attention to the artisan's extraordinary skills, which like the dao can be sensed but not expressed in words. So the Zhuangzi's works contain some intriguing stories describing artisans who demonstrate fascinating artistry, such as butchers, boatmen, wheelwrights, stonemasons, and arrow makers. The fable of butcher Ding cutting up an ox is a well-known example. Zhuangzi used butcher Ding's explanation to King Hui of Wei about how he could cut up an ox so skillfully as to reveal the dao in daily life. As butcher Ding explains,

I have always devoted myself to dao. It is connected to skill. When I first began to cut up bullocks, I saw before me simply whole bullocks. After three years' practice, I saw no more whole animals. And now I work with my mind and not with my eye. When my senses bid me stop, but my mind urges me on, I fall back upon eternal principles. I follow such openings or cavities as there may be, according to the natural constitution of the animal. I do not attempt to cut through joints: still less through large bones.

A good cook changes his chopper once a year—because he cuts. An ordinary cook, once a month—because he hacks. But I have had this chopper nineteen years, and although I have cut up many thousand bullocks, its edge is as if fresh from the whetstone. For at the joints there are always interstices, and the edge of a chopper being without thickness, it remains only to insert that which is without thickness into such an interstice. By these means the interstice will be enlarged, and the blade will find plenty of room. It is thus that I have kept my chopper for nineteen years as though fresh from the whetstone. Nevertheless, when I come upon a hard part where the blade meets with a difficulty, I am all caution. I fix my eye on it. I stay my hand, and gently apply my blade, until with a hwah the part yields like earth crumbling to the ground. Then I take out my chopper, and stand up, and look around, and pause, until with an air of triumph I wipe my chopper and put it carefully away. (Zhuangzi, 1889, pp. 34–35, translation adapted)

The dao in butcher Ding's explanation is a kind of knowledge that is difficult to transfer to another person by spoken or written words; it is tacit knowledge. Zhuangzi's paean to the ancient artisans and their artistry relates to his and other Daoist's advocation of intuition over conscious rationality. At the same time, Zhuangzi clearly also called attention to the inherent value of technics.

Another text from almost two thousand years later attests to continuity in traditional Chinese perspectives on technics: the *Tiangong kaiwu* (Exploitation of the Works of Nature), written in 1607 by Song Yingxing. This work provides detailed descriptions of early Chinese production processes for nearly thirty different fields of agriculture and craft production. The descriptions are further based on firsthand experience with an exceptional range and depth. In comparison with the most important work on technics from the same time period in the West—that is, Georgius Agricola's *De re metallica* (On the Nature of Metals), published in 1556, which only focused on mining—the *Tiangong kaiwu* is remarkable for covering both agriculture and the craft industry. Comparing Song to the famous French encyclopedist Denis Diderot, Needham called him the "Chinese Diderot" (Needham, 1969, p. 102).

The very name of the *Tiangong kaiwu* also deserves comment. It combines the term *tiangong*, meaning "human being replaces Heaven to perform responsibility" from the *Shujing* (Book of Documents), and *kaiwu*, meaning "one succeeds in doing something when he knows and obey the laws" from the *Yijing* (Book of Changes). But Song Yingxing gave this combination new meaning. According to the explanation of Japanese historian of science Saigusa Hiroto, "Tiangong' refers to human behavior which is opposed to nature, "kaiwu" means that human beings transform according to their living interests everything originally contained in nature" (Song, 1992, p. 17).

In Song's view, nature is rich in inexhaustible and precious resources that are not easily acquired from heaven. The only way to attain them is to use manual labor and technics. In other words, as Pan Jixing, the editor of this classic text, has commented, the *Tiangong kaiwu* emphasizes a harmony between heaven or nature and human beings and interactions between human behavior (manual labor) and natural behavior (natural power; Song, 1992, p. 17).

On the basis of his personal experience, Song Yingxing thought it was dishonorable for scholars to have no understanding of where food comes from or how clothes are made, and they should do more than simply immerse themselves in the so-called knowledge of the *Sishu* (four ancient Confucian texts) and *Wujing* (five ancient Chinese classics). Accordingly, Song shifted his attention to practical work (i.e., artisan's work). To collect documents, he traveled the whole country and visited artisans and workers who were actually producing things. In addition, he collected 123 illustrations and carefully structured the 18 chapters of his book, which begins with a chapter on cereals and ends with one on jewelry. Song was prioritizing the necessities of cereals and grain as more important than the luxuries of gold and jade.

This idea of the priority of food over gold and jade has also appeared in other ancient Chinese works, such as the agricultural text *Qi min yao shu* (Main Technics for the Welfare of the People), written by the Northern Wei dynasty official Jia Sixie between 533 and 544 BCE. Even now, Chinese philosophers often attach great importance to the practicality of technics and technologies, especially those related to the Chinese economy and the people's livelihood.

PHILOSOPHY AND TECHNOLOGY IN MODERN CHINA

With the eastward spread of Western culture since the late Ming dynasty (1368–1644) and early Qing dynasty (1644–1911), Chinese reflection began to emerge on the relationship between Eastern and Western cultures. From the 1840s in the late Qing, this reflection began to pay special attention to science and that modern form of technics known as technology.

The immediate stimulus for the modern reflection on science and technology—science being understood primarily in terms of its technological implications—was the defeat of China in the Opium Wars (first in1839–1842 and then again in 1856–1860). As a result, Chinese intellectuals such as Lin Zexu (1785–1850) and Wei Yuan (1794–1857), who were deeply concerned with the crisis facing China since the early nineteenth century, proposed the principle of "learning advanced technology from barbarians in order to oppose barbarians"—that is, importing advanced technology from the West, particularly the technologies of weapons and warships, in order to resist European and American aggression. Chinese civilization was believed to be superior to that of the Western "barbarians." But in order to defend this superiority, China would need to use the technologies of foreigners to be able to resist Western exploitation and imperialism.

In order to fulfill the goal of strengthening itself against the West, the late Qing dynasty accepted this principle and attempted to implement it in a series of institutional reforms associated with the self-strengthening movement. Initially, there were only attempts to import modern weapons. It quickly became apparent, however, that technology transfer by the simple purchase of military technologies was insufficient. Led by Zhang Zhidong (1837–1909), one of the "four famous officials of the Late Qing," the self-strengthening movement was deepened to include "Chinese learning of fundamental principles and Western learning for practical application" (Zhang, 1901).

This expanded approach to technology transfer was summarized in Zhang's *Quan xue pian* (Essay on the Enhancement of Learning) of 1898, a book translated into English in 1901 as *The Only Hope of China*. In this book, Zhang placed his hope in two things: a renaissance of Confucianism and the adoption of Western science and methods. In his words, "The old and new must both be taught; by the old is meant the Four Books, the Five Classics, history, government, and geography of China; by the new, Western government, science, and history. Both are imperative, but we repeat that the old is to form the basis and the new is for practical purposes" (Zhang, 1901, pp. 100–101).For Zhang, China should not just import military technologies from the West but also strengthen itself through the selective adoption of Western education and institutions.

Zhang's position was strongly supported by many elements in the late Qing but was ultimately weakened by a conservative reaction led by the Empress Dowager Cixi (1835–1908). Yet with the creation of the republic in 1912, some version of his policy became the dominant position and has continued into the present. The adoption and application of Western science and technology have been core principles in promoting Chinese industrial development not just in regard to military technology but also with respect to the technologies of mining, steelmaking, textile manufacturing, and so on (Wang, 1994, p. 19). During the nearly one hundred years from the middle of the nineteenth century to the late twentieth century, a strong need to "resist Western aggression (Wang, 1994, p. 20)" drove the Chinese attempt to develop a perspective on technology that would strengthen the nation and increase its wealth.

It was in this historical and cultural context that a special form of Western philosophy—namely, Marxism—was introduced into China and, by the middle of the twentieth century, came to play a central role in Chinese intellectual life. From the founding of the new China—that is, the People's Republic of China (PRC)—on October 1, 1949, until the death of Mao Zedong in 1976, Marxism became a primary influence on Chinese attitudes toward technology.

Although Marxist ideology is not the same kind of educational foundation that Confucian philosophy once was, even since the death of Mao it has remained prominent at every level, from the primary grades to graduate school. Like Confucianism, Marxism also stresses the importance of practice over theory, as when Karl Marx wrote, in the "Theses on Feuerbach" (1845),

> The question whether objective truth can be attributed to human thinking is not a question of theory but is a practical question. (Thesis 2)
>
> All social life is essentially practical. All mysteries which lead theory to mysticism find their rational solution in human practice and in the comprehension of this practice. (Thesis 8)
>
> The philosophers have only interpreted the world, in various ways; the point is to change it. (Thesis 11) (Engels, 1976, pp. 61, 64, and 65)

Each of these theses can be interpreted to echo approaches found in traditional Chinese thinking, although in more radical forms. Indeed, it may be that one factor contributing to the reception of Marxism in China was the practical orientation of Confucian thought that preceded it.

In this respect, it may be useful to note some key respects in which Mao transformed Marxism in the Chinese context. Mao's most philosophical works derive from the late 1930s (after the Long March) and prior to establishment of the PRC. Mao's classics from this period include "On Practice," "On Contradictions," and "Dialectical Materialism," all of which argue for the primacy of experience in producing knowledge (a basic Marxist epistemology with some similarities to American pragmatism) and the revolutionary role of the peasantry (Mao's special contribution to Marxist theory). In "On Practice," Mao paid homage to the importance of technology as a basis for both the production and dissemination of knowledge. As Mao put it, "human knowledge is verified only when a person achieves anticipated results in the process of social practice (material production, class struggle or scientific experiment)" (2009, p. 22). In the first conference on science sponsored by the Communist Party of China (CPC) in 1940, Mao further stated that "the natural sciences [in which he would have included technology] are going to change the world under the guidance of social sciences" (1993, p. 269). Here "social sciences" obviously refers to Marxism. In his opinion, Marxism thus was serving as the foundation for the promotion and development of science and technology.

Along with this perspective, Mao further adopted the Leninist-Stalinist view that philosophy is not an independent activity but a tool of the CPC and correspondingly established institutions for enforcing party orthodoxy. An Australian scholar has summarized the core of ideological Maoist Marxism this way: "By the 1960s Chinese Marxism had become a formalized dogma made up of borrowings from Soviet dialectical materialism, and Mao's formulas, which were more voluntarist. This 'orthodox schema' was the complex result of efforts to identify and propound a Chinese Marxism distinct from its opponents, its local doctrinal variants, and false forms" (Kelly, 2003, p. 435). It was the notion of the revolutionary character of the peasantry and the voluntarist element in Mao's thought that justified the effort of the Great Leap Forward (1958–1961). A distorted emphasis on the importance of experience in knowledge production could also be interpreted as providing some justification for the Cultural Revolution (1966–1976). At the same time, in light of the fact that both of these initiatives caused disasters, some scholars have argued that they did not pay sufficient attention to practice.

Also reflective of the Marxist commitment to science and technology was the proclamation of the Four Modernizations as the basic state goal by PRC Premier Zhou Enlai. First announced in 1954 at the first National People's Congress, the Four Modernizations program was interrupted by the Great Leap Forward's emphasis on peasant power. Following the failure of the Great Leap Forward, in 1963 Zhou reiterated a national need to undertake modernization and strengthening of the fields of agriculture, industry,

national defense, and science and technology. Although the Four Modernizations program seemed to place science and technology on an equal footing
with agriculture, industry, and national defense, in fact it promoted modernization in these three areas through the application of science and technology. The program achieved one of its most obvious and immediate successes
in the 1964 explosion of the first Chinese atomic bomb, followed three years
later by the explosion of the first Chinese hydrogen bomb, followed in 1970
by the launch of the first Chinese satellite (thus realizing the "Two Bombs
and One Satellite" program). But the Four Modernizations were also manifested in ongoing efforts to mechanize agriculture and to create a transportation and industrial infrastructure. After the death of Mao, in 1978 Deng
Xiaoping once again reiterated a commitment to the Four Modernizations,
which became a key feature of the Reform and Opening Up.

A less dogmatic and ideological version of Chinese Marxism, but one that
continues the positive emphasis on science and technology, is associated with
the emergence of what is called "dialectics of nature" research. The term
derives from a book of this title by Friedrich Engels (1820–1895), but in
China, it has come to include discourse that in the West is called history
and philosophy of science and technology. When dialectics of nature discourse emerged in the late 1950s, it did not initially attract much attention.
Eventually, however, it gained acceptance among scientists, engineers, and
philosophers as a framework within which critical reflection could take place.
As stated in the National Twelve-Year (1956–1967) Plan of Science and
Technology Development, "just as historical materialism mediates between
philosophy and social science so there exists a form of knowledge between
philosophy and natural science. Because this [mediation between philosophy and natural science] continues the research initiated by Engels in The
Dialectics of Nature, we call it Dialectics of Nature . . . This development
emphasizes cooperation between philosophers and natural scientists" (Gong,
1991, p. 4).

Under the dialectics of nature rubric, philosophers and scientists formed a
close alliance both theoretically and practically. Because technological innovation and revolution had become such major commitments in the PRC,
studies of technological innovation from an epistemic point of view also
acquired importance in philosophy. During 1958–1960, some veteran workers at a factory in the northeast industrial city of Harbin became leaders
in the invention of some heavy equipment machine tools. A research team
composed of philosophical and technological scholars from the Harbin Institute of Technology visited the factory in order to study the features of this
successful work. They concluded that success could be expressed in Marxist
terms as "grasping the principal contradiction" in any particular situation.
Moreover, the scholars argued, others "should learn from them the experience of analyzing the problems in our 'scientific experiment' through the
point and method of material dialectics in order to promote the learning and
research on dialectics of nature" (Guan, 2000, p. 18). According to one later
commentator, this approach "greatly attracted the interest of the national

academic field of mechanical engineering. It generated heated discussions among a number of university, factory, and institute scholars. It initiated learning, research, and application of the dialectics of nature in the field of engineering technology" (Zhou, 1983, p. 113).

Although the alliance between philosophers and engineers was interrupted during the Cultural Revolution, it was revived after the Reform and Opening Up in order to promote the national task of socialist modernization. In 1981, the first annual meeting of the Chinese Society for Dialectics of Nature (whose establishment was approved by Deng Xiaoping) "mostly analyzed and discussed how dialectics of nature work could serve to promote the construction of China's socialist modernization in a new historical stage." A strong belief was expressed that the "dialectics of nature could exert its unique function" in this new historical period (Huang and Zhou, 1988, p. 411).

Chinese philosophy of technology was carried forward under the dialectics of nature rubric until the Academic Degrees Committee of the State Council in 1987 renamed the field the philosophy of science and technology. During the 1980s, some important philosophical works published in the area included *Kexue jishu lun* (Theory of Science and Technology; Yang Peiting et al., 1985), *Kexue jishu xue* (Science of Science and Technology; Meng Xianjun et al., 1986), *Lun jishu* (On Technology; Yuan Deyu et al., 1986), *Jishu lun* (Theory of Technology; Chen Nianwen et al., 1987), and *Jishuxue daolun* (Introduction to Technological Science; Deng Shuzeng et al., 1987). In all these cases, there was an effort to distinguish technology from science so that technology could be accorded a proper emphasis in education and government funding. In addition, a number of related monographic studies were published in Chinese at different technological universities: "On the Methodology of Engineering Technology" (Northeast Technical College), "On the Methodology of Technology Development" (Dalian Technical College), "On the Structure and Development of Engineering Technology" (Harbin Institute of Technology), and "Japanese Philosophy of Technology" (Chengdu University of Science and Technology). In these, it is possible to identify an emerging concern with the more effective management of engineering and technology. Finally, on the basis of a general theory of technology, efforts were made to clarify the character of professional work in such specific disciplines as chemical engineering, petroleum engineering, agricultural engineering, biomedical engineering, and military engineering.

The increasing number and size of engineering construction projects in recent decades have shifted philosophical attention from technology to engineering. In this sense, the quest for a proper appreciation of technology has become a quest for a general appreciation of engineering. Work in the philosophy of engineering was stimulated especially by the post-Mao leadership of Deng Xiaoping, who in 1988, furthering Zhou Enlai's promotion of the Four Modernizations, declared science and technology as primary productive forces, which in reality take place through engineering. Also worth noting is the fact that discourse in the philosophy of engineering, which again stressed its positive aspects and the ability of engineering to contribute to economic and social development, has found a receptive home in the Chinese Academy of Engineering (CAE). As a new branch of philosophy, philosophy of engineering reveals a strong practical aspect, related to but different than philosophy of technology: "Philosophy of engineering is the overall consideration of engineering activity, which is the human activity of depending on, adapting, knowing and changing nature, and the knowledge of fundamental principles and universal laws concerning engineering activity . . . The only way to promote its development is to approach engineering practice through philosophical reflection" (Yin et al, 2007, p. 16).

According to an interpretation advanced by Wu Guosheng, Western philosophy has from its birth paid more attention to theory than to practice (Wu, 1999). Since the early twentieth century, some philosophical schools in the West have challenged this emphasis. To some degree, the philosophy of technology and engineering are involved in this challenge and can thus be seen as scholarly approaches with a vital future insofar as they manifest a strong tendency to closely link theory with practice. As another scholar has summarized the situation in China: "It deserves to be appreciated that the philosophy of engineering insists on the point of view of practical philosophy, considers the principle of relating theory with practice, and emphasizes the dialogue between philosophers and engineers and other engineering practitioners. Precisely in this way can it have bright future" (Yuan, 2007, p. 111).

It should be emphasized that this is a very limited description of the development of philosophical perspectives on China in the contemporary period. There is at present a significant flowering of philosophical work on technology and engineering that has not been mentioned. The aim here has only been to present some of the general background.

CONCLUSION

This chapter has argued that technological literacy in China manifests a more positive appreciation of technology than is often the case in the developed world. Such an appreciation is evident even in traditional Chinese philosophy, where practical affairs receive more attention than in premodern Western philosophy. Moreover, there is a tradition of scholars actually writing about technics that is much older and more pronounced than in the West. It is this that helps explain what Needham has described as the unequaled technical inventiveness in China for more than two thousand years until roughly 1500. Since the foundation of the new China (1949 to the present), first Marxist theory and then the Reform and Opening Up movement have revived and intensified the Chinese appreciation of technology. Technology in China is seen as fundamentally good because it both reduces the burden of human labor and increases human productivity. This is an assessment of which the more technologically developed West should not totally lose sight. Nevertheless, although in China today the many unintended consequences of technological development have not yet overwhelmed positive assessments, such experiences as environmental pollution are beginning to stimulate some qualifications.

In addition, there are three key ideas that, although not examined in the present chapter, have been central influences on Chinese technological literacy and would thus deserve careful attention in a more extended discussion. First, there is the absence of a creator god (or, more positively, the sense of the material world as self-subsisting). There is nothing like the almighty or supreme God or Allah, as found in Western thought, who creates the whole world, including human beings, out of nothing. On the contrary, in Chinese thought, although creation and destruction go on in the world (in cyclical patterns), the world as a whole cannot itself be created. The sages, or rather human beings, create all things according to universal principles. Second, there is the emphasis on practice (and the primacy of practical or political affairs in human life), which has been discussed to some extent. On the one hand, this involves the promotion of "practical rationality." As Li Zehou has written, "human rationality in ancient China . . . tended toward practical research that would help people obtain useful knowledge for their lives" (Kim, 2012, p. 282). On the other hand, a practice orientation in the Chinese context places an emphasis not just on economic values such as high efficiency, increased profits, and reduced costs but also on social goods such as familial bonding, shared prosperity, and a peaceful life for the people. Finally, there is a concern for harmony between heaven and earth (i.e., of human beings with the larger world in which they live). Humans and nature are considered as a whole, and a harmonious relationship between the parts (humans) and the whole (heaven or nature) is always the focus of thinking—specifically, attempting to achieve *tian ren he yi* (the unity of the heaven and humans). The future of a distinctive Chinese technological literacy can be expected to involve the further development of these three key ideas against the historical and philosophical background that has been sketched here.

ACKNOWLEDGMENTS

This chapter is adapted from a previous article: "Philosophical Perspectives on Technology in Chinese Society," *Technology in Society* 35, no. 3 (2013): 165–71. The author also wishes to acknowledge Carl Mitcham's assistance in the final English editing and credit support for related research to a postdoctoral fellowship from the Hennebach Program in the Humanities at the Colorado School of Mines.

REFERENCES

General note: In all quotations, I have silently substituted pinyin for any other transliteration system. From all Chinese sources, the translations are my own.

Chen, Guying. (1999). *Zhuangzi qian shuo* (The brief introduction on Zhuangzi). Beijing, China: Zhonghua shuju (Zhonghua Book Company).

Cua, A.S. (2003). Philosophy in China: Historiography. In A.S. Cua (Ed.), *Encyclopedia of Chinese philosophy* (pp. 499–510). New York, NY: Routledge.

Engels, Frederick. (1976). *Ludwig Feuerbach and the end of classical German philosophy*. Beijing, China: Foreign Language Press.

Feng, Youlan. (2007). *Zhongguo zhexue jian shi* (A short history of Chinese philosophy). Bilingual English and Chinese edition. Tianjin, China: Tianjin shehui kexue chubanshe (Tianjin Academy of Social Sciences Press).

Gong, Yuzhi. (1991). Ziranbianzhengfa shi (History of Chinese dialects of nature). *Ziranbianzhengfa yanjiu (Studies in Dialectics of Nature)*, 7(2), 1–5.

Guan, Shixu. (2000). Li Chang yu Hagongda ziranbianzhengfa yanjiu (Li Chang and studies on dialectics of nature by Harbin Institute of Technology). *Ha'erbingongyedaxue xuebao (Journal of HIT, Social Sciences Edition)*, 2(2), 16–24.

Huang, Shunji, & Zhou, Ji. (1988). *Ziranbianzhengfa Fazhan shi* (The Development History of Dialectics of Nature). Beijing, China: Zhongguorenmindaxue chubanshe (Chinese Renmin University Press).

Kelly, David. (2003). Marxism in China. In A.S. Cua (Ed.), *Encyclopedia of Chinese philosophy* (pp. 431–438). New York, NY: Routledge.

Kim, Hongkyung. (2012). *The old master: A syncretic reading of the Laozi from the Mawangdui text Aonward*. Albany, NY: State University of New York Press.

Kongzi. (2003). *Confucius Analects: With selections from traditional commentaries*. Trans. Edward Slingerland. Indianapolis, IN: Hackett.

Liu, Juncan. (1996, August). Zhongguo chuantong jishu sixiang yu xifang sixiang de duizhao (A comparison between ancient Chinese technological thoughts and Western ones). *Kexue yuekan (Science Monthly)*, 320, 686–690.

Mao, Zedong. (1993). *Mao Zedong wenji* (Mao Zedong's collected works). Zhonggongzhongyang wenxian yanjiushi (Party Documents Research Office of the CPC Central Committee) (Eds.). Beijing, China: Renmin chubanshe (People's Publishing House).

Mao, Zedong. (2009). *Collected writings of Chairman Mao*. Vol. 3, *On policy, practice and contradiction*. Shawn Conners (Ed.). Foreign Language Press (Beijing) (Trans.). El Paso, TX: El Paso Norte Press.

Munro, Donald J. (2001). *The concept of man in early China*. Ann Arbor: University of Michigan Press.

Needham, Joseph. (1969). *The grand titration: Science and society in East and West*. Toronto, Canada: University of Toronto Press.

Needham, Joseph, &Wang, Ling. (1963). *Science and civilization in China*. Vol. 4, *Physics and physical technology*, Part II, Mechanical engineering. Cambridge, UK: Cambridge University Press.

Song, Yingxing. (1992). *Tiangong kaiwu yizhu* (The translation and annotation of *Tiangong Kaiwu*). Annotated by JixingPan. Shanghai, China: Shanghai guji chubanshe (Shanghai Classic Publishing House).

Sun, Bin. (2010). Lun Xi Zhou cong "philosophoy" dao "zhexue" yici de fanyi guocheng (Discussion of the translation of philosophy into Zhexue by Nishi Amane). *Qinghuadaxue xuebao, zhexue yu shehui kexue ban (Journal of Tsinghua University, Philosophy and Social Science)*, 25(5), 122–131.

Wang, Ermin. (1994). *Zhongguo jindai sixiangshi* (History of modern Chinese thought).Taipei, Taiwan: Taiwan shangwu yinshuguan (Taiwan Commercial Press).

Wen, Renjun. (1993). *Kaogongji yi zhu* (Translation and annotation of Kaogongji). Shanghai, China: Shanghai guji chubanshe (Shanghai Classic Publishing House).

Wu, Guosheng. (1999, November 17). Jishuzhexue: yimen youzhe weida weilai de xueke (Philosophy of technology: A scholarly discipline with a brilliant future). *Zhonghua dushu bao (China Reading Weekly)*.

Yang, Shengbiao, & Xu, Kang. (2002). Gongcheng fanchou yanbian kao lue (Evolvement of the category of the engineering). *Ziranbianzhengfa yanjiu (Studies in Dialectics of Nature), 18*(1), 38–40.

Yang, Shuda. (1954). *Jiweiju xiaoxue shu lin* (The primary studies and comments written at the residence of Jiwei). Beijing, China: Kexue chubanshe (Science Press).

Yin, Ruiyu, Wang, Yingluo, &Li, Bocong (Eds.). (2007). *Gongcheng Zhexue* (Philosophy of engineering). Beijing, China: Gaodengjiaoyu chubanshe (Higher Education Press).

Yuan, Deyu. (2007). Cong jishuzhexue de shijiao kan gongchengzhexue (On philosophy of engineering from the angle of the philosophy of technology). *Ziranbianzhengfa yanjiu (Studies in Dialectics of Nature), 23*(12), 110–112.

Zhang, Zhidong. (1901). *China's only hope.* Trans. S. I. Woodbridge. Edinburgh, Scotland: Oliphant, Anderson and Ferrier.

Zhou, Lin (Ed.). (1983). *Zhongguo ziranbianzhengfa yanjiu lishi yu xianzhuang* (The history and current situation of Chinese dialectics of nature). Beijing, China: Zhishi chubanshe (Knowledge Press).

Zhuangzi. (1889). *Zhuangzi: Mystic, moralist, and social reformer.* Trans. Herbert A. Giles. London, UK: Bernard Quaritch.

EDUCATION AND THE CONCEPT OF
TECHNOLOGICAL LITERACY

CHAPTER 8

ENABLING BOTH
REFLECTION AND ACTION

A CHALLENGE FACING
TECHNOLOGY EDUCATION

David Barlex

This chapter argues that reflection and action have been fundamental to design and technology education in England since its inception in 1988. It describes and exemplifies a pedagogy that can be used in different areas of study within the subject. The chapter then describes recent developments involving the revision of design and technology within the National Curriculum in England and their possible relationship to reflection and action. Finally, the chapter discusses the challenge of embedding pedagogy for reflection and action in authentic tasks and emphasizes the importance of any statutory requirements, including both action and reflection as key components of the subject.

JUSTIFYING REFLECTION AND ACTION
AT A TIME OF UNCERTAINTY

In England at the time of writing, the government had instigated a review of the National Curriculum and commissioned an expert panel to provide detailed advice on the construction and content of the new National Curriculum (Department for Education 2011). From the beginning of this process, some subjects were privileged above others. English, mathematics, science, and physical education were to be included in the new curriculum, and the status and provision of statutory programs of study for all other subjects were to be decided. The expert panel identified three curriculum areas:

- The National Curriculum, a combination of core and foundation subjects in which the core subjects are specified for each key stage through detailed programs of study and attainment targets and foundation subjects are specified for each relevant key stage through significant but refined and condensed specifications.
- A basic curriculum that describes the statutory requirements for curricular provision in addition to the National Curriculum. These are compulsory requirements, but schools are able to determine for themselves the specific nature of this provision and the government will not produce programs of study and attainment targets for subjects and topics in the basic curriculum.
- A local curriculum that allows substantial scope for curricular provision determined at the school or community level. This might include additional subjects or courses, but it should also enable schools to extend or contextualize the national and basic curricula in ways best suited to the needs of particular groups of pupils.

The expert panel recommended that design and technology should be reclassified as part of the basic curriculum on the grounds that, although it is important in balanced educational provision, it did not have sufficient disciplinary coherence to be stated as a discrete and separate National Curriculum subject. The argument was that design and technology had weaker epistemological roots than those categorized as core or foundation subjects.

There can be little doubt that the subject of design and technology is broad, requiring pupils to work across a wide range of materials and associated fields—food, textiles, graphic media, resistant materials, control systems, and new and smart materials—and adopt a design-based approach using both conventional and digital methods. This breadth, plus the recent criticisms from Ofsted (2011), paints a picture that would convince those not committed to the subject being an essential and clearly defined part of a National Curriculum that it lacked the established orthodoxy enjoyed by, for example, school subjects such as mathematics and science. "However, most of the schools visited had not made sufficient use of subject-specific training to enable teachers new to the profession, and those who were more experienced, to continually update their subject knowledge. This often resulted in an outdated Key Stage 3 curriculum, an issue that also related to the upper end of Key Stage 2. In around a third of the secondary schools, too little use was made of electronics, computer aided design and manufacture (CAD and CAM) and control technology in the teaching of D&T" (p. 5).

At the time of writing, Minister for Education Michael Gove had yet to respond to the expert panel's recommendations, and it is against this backdrop of inevitable but uncertain change that this chapter will consider the requirements for both action and reflection within the school subject of design and technology. Although lacking the disciplinary coherence of subjects such as mathematics and science, design and technology is unique in that it teaches young people "to take action." The interim report (Department for Education and Science, 1988) that preceded the introduction of design and technology in the original National Curriculum captured this well in response to the question "What is it that pupils learn from design & technological activities which can be learned in no other way?" In its most general form, the answer to this question is that students learn to operate effectively and creatively in the made world. The goal is increased "competence in the indeterminate zones of practice."

Jacob Bronowski (1973), in his brilliant book *The Ascent of Man*, considered this ability to take action as a defining feature of humanity. In his view, the actions taken by humankind are not like those of other animals.

Among the multitude of animals that scamper, fly, burrow and swim around us, man is the only one who is not locked into his environment. His imagination, reason, emotional subtlety, and toughness make it possible for him not to accept the environment but to change it (p. 19).

The tool that extends the human hand is also an instrument of vision. It reveals the structure of things and makes it possible to put them together in new, imaginative combinations (p. 120).

Underpinning Bronwoski's vision of the uniqueness of humanity is the power to imagine and conceive new possibilities that might be achieved through taking action. Without this reflection, there can be no purposeful action, and a challenge facing technology education is to ensure an approach to teaching technology that embraces both action and reflection such that they are in a synergic relationship.

Pedagogy for Reflection and Action

The pedagogy described here was devised with the intention of achieving synergy between action and reflection. It derives from the activities of three curriculum projects: Nuffield Design and Technology (Barlex, 1998; Givens & Barlex, 2001), Young Foresight (Barlex, 2001, 2003), and Electronics in Schools (Barlex, 2004). Each of these curriculum projects involved a dynamic relationship between curriculum development and research. The Nuffield Design and Technology Project called on the work of the Assessment of Performance Unit in Design and Technology (Barlex &Welch, 2001). Two independent evaluation exercises informed the final form of the Young Foresight project (Murphy et al., 2000, 2001). The independent evaluation of Electronics in Schools (EIS) has validated the approach to design-and-technology professional development initially elaborated through international research from four different countries: England, Canada, Finland, and New Zealand (Banks et al., 2004).

In the period from 1990 to 1995, the Nuffield Design and Technology Project developed three types of learning activities: capability tasks, resource tasks, and case studies. A capability task requires pupils to design and make products that work and in so doing learn to become capable and reveal their capability. This involves pupils in generating and developing design ideas and combining them into a visualization of the product so clear that they can communicate it through sketches, drawings, and three-dimensional models that they can use as a guide when making the product. This product will embody their design ideas and reveal both their designing and making abilities.

If pupils are not reflective when tackling a capability task, the action they take will be poorly conceived and lead to disappointing and inappropriate outcomes. Such tasks are engaging but highly demanding activities, and pupils are unlikely to be successful if they cannot build on prior learning. Resource tasks and case studies provide this prior learning. A resource task is a short activity, often practical, that requires pupils to think and helps them learn the knowledge and skills they need to design and make products really well—that is, to be successful in a capability task. Through resource tasks, pupils learn design strategies, communication techniques, making/manufacturing skills, technical knowledge and understanding, and commercial matters. The range of learning encompassed by resource tasks supports both reflection and action. A case study is a true story about design and technology from the world outside school. Through case studies, pupils learn how firms and business design and manufacture goods and how goods are marketed and sold. Pupils also learn about the impact that products have on the people who use them and the places where they are made and used. These impacts can be economic, social, or environmental and may be unanticipated by both producer and user. Learning from case studies primarily supports reflection. In both resource tasks and capability tasks, the emphasis is on individual learning and performance. However, in case studies, there are sometimes instructions requiring pupils to work collaboratively and discuss the contents of the study with a partner.

The Young Foresight project, which began in 1997, adapted the Nuffield capability tasks and resource tasks in four important ways. First, it removed the requirement to make from the capability task so that pupils could concentrate on designing without being limited by their personal making skills and the facilities available in the school. Second, it required pupils to work collaboratively in this design activity. Third, it reconceptualized the supporting resource tasks into a toolkit of tasks to support designing rather than designing and making. Fourth, it required pupils to discuss their work with one another when tackling toolkit tasks. The emphasis on collaboration and communication is in considerable contrast to the mainly individualistic way in which pupils were expected to work within the Nuffield approach to design and technology. Here, taking action was limited to making design proposals, but reflection was given a high priority in that pupils were required to decide for themselves what they would design and to justify its worth.

The emphasis on design *and* making in the Nuffield Design and Technology Project was maintained in the EIS project, which began in 2001, but the design decisions made by the pupils were made more explicit by considering them as belonging to one of five sets: conceptual, technical, aesthetic, constructional, and marketing. These sets of design decisions are mutually interdependent. A change of decision in one set will almost certainly lead to changes in the other sets. By making the possible design decisions more explicit, the EIS project was able to provide teachers with an audit tool by which they could scrutinize the range of designing required in capability tasks and ensure that pupils were properly prepared to make and justify these decisions. By clarifying and focusing on the design decisions that will be enacted through making, this approach requires a high degree of reflection in making the design decisions under the imperative that they must be achievable through making. Hence both reflection and action are given a high priority.

ACHIEVING ACTION AND REFLECTION: CONSIDERING HOW NEEDS AND WANTS ARE MET BY TECHNOLOGICAL PRODUCTS AND SERVICES

If pupils are to move away from an egocentric approach to designing, it is important that they design for others rather than for themselves. To do this, they will need to have some understanding of people's needs and how these might be met by products and services. The Nuffield Design and Technology Project developed the PIES approach to identifying human needs. PIES is an acronym standing for physical, intellectual, emotional, and social and provides a simple conceptual framework for classifying needs. We need food, water, and air to breathe. We need to keep warm, be protected from the weather, and exercise regularly. These are physical needs. We need to be mentally active, learn new things, and be stimulated. These are intellectual needs. We need to feel safe and secure. We need to feel that others care about us and to have ways of expressing our feelings. These are emotional needs. Most people like to spend time with their friends, talking and sharing joint activities. These are social needs. The resource task that introduces this approach asks pupils to identify the needs of people in three different situations: in a hospital ward, in a hotel room, and on a train journey. It then requires the pupils to identify products that meet these needs. In this way, the pupils can learn about the relationship between needs and products. It provides pupils with a conceptual tool that they can use to interrogate both situations and products. They can ask of situations they observe, "What are the needs of people in this situation?" and "What products do they use to meet these needs?" They can ask of existing products, "What needs will be met by this product?" and "In what situations are people likely to have these needs and hence use these products?" They can put their own product-design ideas into a variety of situations and scrutinize them for their potential usefulness. By using the PIES approach to considering needs and wants, teachers can develop in pupils a user-centered approach to designing and making so that the actions they take through designing and making are informed by significant reflection on peoples' needs and wants.

The Young Foresight toolkit introduces the PIES approach through considering the needs of people who are waiting at a bus stop, a railway station, and an airport and identifying the needs met by a mobile phone, laptop computer, and newspaper. This is then extended by means of a second activity in which pupils explore the differences between needs and wants. This is done by means of two short vignettes that describe how the needs of people in very different situations are met. This leads pupils to appreciate that what we want and can hope to acquire to meet our needs are determined largely by the nature of the society in which we live. Pupils' actions are limited here to the extent that they will devise design proposals that they will not make, but their proposals will be informed by a consideration of users' needs and wants, ensuring that both action and reflection are required.

Several of the case studies in the *Nuffield Design and Technology Study Guide* (Barlex, 1995a) reinforce the relationship between people's needs and the products and technologies they use. The case study "Printing—from wood-blocks to computers" illustrates how technological advances in printing have been used to meet changing needs. "Designing houses to suit people's needs" describes housing projects in Peru that illustrate how important it is to involve local people in making design decisions. "Designing maths instruments for different users" describes how a compass was redesigned to suit the needs of young children. By regularly requiring pupils to read and discuss such case studies, teachers can reinforce understanding of the important relationship among needs, wants, and the nature of available products. It is important that teachers help pupils utilize this learning from case studies, which is essentially reflective, to inform the actions they take when designing and making.

Considering Design Decisions as the
Basis for Design and Make Activities

The EIS project was a professional development initiative, and during the ongoing evaluation that took place during its first year of operation, it was realized that the majority of the in-service training taking place was concerned with technical subject knowledge about electronics. This is not surprising, but dealing with these technical issues, while important, was not sufficient to ensure that there was an impact on pupils' learning in design and technology lessons. It was important to focus on the totality of electronic product design and consider the wide range of design decisions, including the technical, that pupils would need to make if they were to be successful in electronic product design. Subsequently, this holistic approach to electronic product design was found to generate tasks that were authentic in that they were personally meaningful and set in culturally authentic contexts (Murphy & Hennessy, 2001) and could be applied across the breadth of a design and technology program of study. The view of the EIS project was that pupils should be given the opportunity to learn to make five types of design decisions:

1. Conceptual (e.g., What is the overall purpose of the design? What sort of product will it be?)
2. Technical (e.g., How will the design work?)
3. Aesthetic (e.g., What will the design look like?)
4. Constructional (e.g., How will the design be put together?)
5. Marketing (e.g., Who is the design for? Where will it be used? Where will it be sold?)

These can be represented visually, with each type of decision at a corner of a pentagon, with each corner connected to every other corner. This interconnectedness is an important feature of design decisions. A change of decision within one area will affect some if not all of the design decisions made within the others. For example, if the way a design is to work is changed, this will almost certainly affect what the design looks like and how it is constructed. It may also have far-reaching effects in changing some of the purposes that the design can meet and who might be able to use it. Usually the teacher identifies the sort of product the pupils will be designing and making. This makes it very difficult for pupils to engage in conceptual design. Even if the type of product is identified for the pupils, there are still many opportunities for making design decisions in the other areas. Consider the designing and making of a puppet theater and puppets. The pupils can make decisions about who will use the puppets and what they will be used for (marketing decisions), what sort of puppets would be appropriate, the sort of theater such puppets would need, the nature of props and scenery, plus any special effects that might accompany the performance. These decisions will encompass a host of technical, aesthetic, and constructional design decisions.

It is through a combination of the learning from appropriate resource tasks and previous experience within capability tasks that pupils will be empowered to make these design decisions. Teachers are able to use the idea of design decisions to scrutinize their design and technology curriculum. The first step is to audit the range of design decisions that are likely to be made by pupils tackling a particular designing-and-making assignment. The second step is to carry out this audit across all the designing-and-making assignments tackled by pupils across a key stage. This gives an overview of the designing that is taking place. If an area of design decision making is missing, underrepresented, or overrepresented, the nature of the assignment can be adjusted accordingly. Teachers can adapt their curriculum to include resource tasks that are relevant to the required design decisions. In this way, the demands for reflection within designing can be orchestrated by the teacher so that pupils are required to make and justify their design decisions in ways that are challenging without being daunting and without being overwhelmed with design tasks that are too complex. As pupils become more adept at this reflection about design decisions, the actions they take as a result will also become more highly developed. As pupils move through a program of study requiring them to make and enact design decisions, they will develop a growing interdependence between reflection and action.

Designing without Making in Response
to New and Emerging Technologies

Insisting that pupils should always make what they have designed can under-mine their autonomy, especially if they have limited making skills. The Young Foresight project deliberately avoids this difficulty by requiring pupils to work collaboratively in designing but *not* making products and services for the future. The Young Foresight approach identifies four factors that teachers should encourage their pupils to take into account:

1. *The technology that is available for use.* This should be a new and/or emerging technology and be concerned primarily with how the new product or service will work. Pupils should not concern themselves with manufacturing.
2. *The society in which the technology will be used.* This will be concerned with the prevailing values of the society: what is thought to be important and worthwhile. This will govern whether a particular application of technol-ogy will be welcomed and supported.
3. *The needs and wants of the people who might use the product or service.* If the product does not meet the needs and wants of a sufficiently large number of people, then it will not be successful.
4. *The market that might exist or could be created for the products or services.* Ideally, the market should one with the potential to grow, one that will last, and one that adapts to engage with developments in technology and changes in society.

Clearly, these factors interact with one another and influence the sorts of products and services that can be developed and will be successful. Using this way of thinking, unencumbered by the necessity of making the proposed designs, enables pupils to be creative and develop highly original, conceptual design proposals.

An important decision for the teacher is the order in which to ask pupils to tackle the task. One way is to start with a particular new technology and ask a sequence of questions such as these: What sorts of things can we use this technology for? What needs would the technology meet? Would meet-ing these needs be seen as important and worthwhile in society in the future? Would people want products or services to meet these needs? What sort of market is there likely to be for these products and services?

If this approach is adopted, it will be important to be wide ranging in answering the first question. For example, in asking this question about the possible application of quantum tunneling composite (QTC), a new material that can be used to make pressure-sensitive conductors, a group of pupils identified the following possible areas of application: sport and leisure, trans-portation, medicine, environmental monitoring, and aids for the handi-capped. Eventually, they focused on producing a range of sport-and-leisure goods and developed two ideas from this range: a device that could be used to help people recover from hand injuries and overcome arthritis and a textile product that could be used for step exercises and would keep an accurate record of exercises performed. They were able to justify both product ideas in their answers to the other questions. And they were able to explain how the technology they had started with would be utilized in their designs.

Another way is to start by asking pupils to construct a scenario of what a future society will be like and what life will be like for particular groups of people in that society. Pupils can then explore a sequence of questions such as these: What needs would there be in that society? What products and services would people want to meet these needs? What sort of market is there likely to be for these products and services? What technology do I need to make the product or service work?

This is a much more demanding approach, but it offers more scope for considering the nature of a future society and the impact of technology on that society. It is an approach that is more likely to stall, as the starting point is much less concrete than a particular technology. However, it does have the potential for developing some really big ideas. For example, a group of pupils constructed a scenario in which only the rich had access to antiaging technology through private medical care. And the government refused to make it available through the National Health Service because this would lead to an increase in demand for health and other social services that would be unsustainable. In this scenario, there was considerable social unrest, and action groups used the Internet to mobilize opposition to the government. The pupils created another scenario in which this technology was available to all and explored ways in which the active elderly could make a financial contribution to society by means of limited work from home using the Internet. The limited work from home involved a wide range of activities using the Internet: providing companionship for the lonely, providing tutorial support for those studying at school using e-learning, supporting a forum for the discussion of local issues to develop participation in local government, providing examples of oral history from their memories of times past, and providing guidance to those involved in work similar to that which they did when in full-time employment.

The design that the pupils produced was a guide to antiaging services provided by new technology. A comparison of the two scenarios revealed that the way a society makes such technology available to its people has consequences and showed how the same technology, in this case use of the Internet, could be used for very different purposes.

A concern expressed by some teachers about the Young Foresight approach was that pupils would become demotivated if they were not able to make their design proposals on an individual basis and have them for themselves. In reality, the reverse was true. The pupils valued the collaborative approach and saw the benefits in comparison with working as isolated individuals.

- You get a better product as a group. You don't get "Ah, you're doing this wrong";

- They help you out and say you could have done this . . . help you evaluate your product sort of.
- You're not like "do this, do that", you've got more ideas to do it. More opportunity.
- Not just using your own ideas, you get other people's and you can work off them.
- It works better when we're in groups, more ideas than if you work as an individual. You can see things from different perspectives.
- It's really shown me what I can achieve, and as a group as well. (3; 4)

They also valued their design ideas, particularly as they themselves were responsible for identifying the design tasks they tackled. Barlex and Trebell (2008) reported that pupils were very keen to share their design ideas, prompting detailed feedback such as the following: "I liked it, as I designed a ring that allows you to download movies and allows you to project them onto a surface to watch them." They were clearly proud of their design ideas, prompting comments such as the following: "I really liked it as it was well thought out, and the product itself was the shape of an eye." They all agreed that the need to make would have led to simplification of the design and that the unit taught them "to design ideas from other peoples' perspectives and wants." So although taking action is limited to developing a design proposal, there is no doubt that this approach requires a high level of reflection on the part of the pupil. The pupils' commitment to their design proposals was significant, indicating that they valued this more highly than items they could actually make.

Understanding Impact beyond Intended Benefit

The brief history of the automobile, invented a little over one hundred years ago, shows that some technologies have far-reaching effects beyond their intended benefits. The development of the internal-combustion engine gave rise to a transport system that completely revolutionized our way of life. To accommodate the needs of the motorist (and to provide for movement of goods by trucks and tankers), a large network of roads and motorways have developed. The use of motor vehicles on this transport network contributes significantly to pollution of the atmosphere and global warming. Learning to drive and acquiring an automobile have become a rite of passage for most young adults, male and female, in many countries. The opportunity to move from your place of birth to new and different places, to gain employment, to meet new people, and to form friendships and relationships are facilitated by the automobile.

This physical and social mobility can have a deleterious effect on small, localized communities. In the United States in 2002, there were an estimated 6,316,000 automobile accidents, resulting in approximately 2.9 million injuries and 43,000 people killed. Since the first crash fatality in 1896, motor vehicles have claimed an estimated 30 million lives globally. On average, someone dies in a motor vehicle crash each minute in the world.

When the automobile was invented and the automobile industry was born, no one envisaged that subsequent design iterations would be responsible for environmental damage, social upheaval, and a colossal death toll. What would the Victorian designers and engineers responsible for developing the automobile have done if it had been suggested to them that their work would harm the planet, erode family values, and kill millions of people? They would probably have been incredulous. They might have argued that society would step in and stop all those dreadful consequences by managing the way this new technology would be used. But they would have been wrong. We know this because these events have come to pass. And of course those Victorian engineers and designers did not have a malignant intent. They saw what they were doing as providing considerable benefits to many people. And in that respect, they were correct. The automobile has been highly beneficial to many individuals, communities, and societies, but at a cost: the cost of impact beyond the intended benefit.

A challenge for the technology curriculum is to engage young people with a product's impact beyond the intended benefit, and a curriculum that limits pupils to designing and making simple products that can be manufactured in school workshops will fail to achieve this. This challenge was addressed by a relatively recent project commissioned by Foresight, a department within the Department for Business, Innovation, and Skills. Entitled "Change of place" (Barlex 2006), it required young people to consider the future of cities with a particular emphasis on intelligent infrastructure. A subtext to the activities in the project was impact beyond intended benefit. In this case, the pupils' task was to ameliorate the negative impacts beyond intended benefit caused by many features of transportation in cities. The project adopted the resource task/capability task pedagogy of the Nuffield Design and Technology Project and set the main tasks shown in Table 8.1.

Clearly the tasks require significant reflection about both the problems facing existing cities and how these might be addressed through either modifying existing cities or devising new ones. The presentation of findings can be seen as taking action (albeit intended action as opposed to actual action). It is important that these intended actions have significance for the pupils and that their presentations of findings are taken seriously by those whose opinions the pupils value. These include their peers, family members, senior leadership in their schools, and members of the wider community outside school, such as local business leaders and local government officials. Inviting individuals from these various stakeholder groups to consider and constructively criticize pupils' presentations is a powerful way of ensuring pupils' learning engages with both reflection and action.

RECENT DEVELOPMENTS IN ENGLAND IN RESPONSE TO THE CURRICULUM REVISION

Education for Engineering (E4E) is the body by which the engineering profession offers coordinated and clear advice on education to the UK

Table 8.1 Tasks in the Change of Place Project

Developing existing cities:

1. Find out what is wrong with cities as they are today.

2. Develop proposals to show how the situation can be improved.

3. Present your ideas for improvement.

Presenting findings in the following ways:

- A PowerPoint presentation describing the changes you would make and why these are an improvement

- A comic strip showing a day in the life of someone living and working in your improved town or city

- A front page of a newspaper describing the improved town or city

- A storyboard of advertisements for living in the improved town or city

Envisioning new cities:

1. Develop a scenario for a new city of the future.

2. Develop proposals to show your vision for a new city of the future.

3. Present your vision of the new city of the future.

Presenting findings in the following ways:

- A PowerPoint presentation describing what is important and how your city deals with this.

- A comic strip showing a day in the life of someone living and working in your new city.

- A front page of a newspaper describing your new city.

- A storyboard of advertisements for coming to live in your new city.

government and the devolved assemblies. It is hosted by the Royal Academy of Engineering with a wide membership drawn from the professional engineering community. In an attempt to respond to the criticism of the expert panel that design and technology did not have sufficient disciplinary coherence to be stated as a discrete and separate National Curriculum subject, the E4E assembled an expert group of design and technology specialists from outstanding schools across England, higher education academics, representatives from industry, the Royal Academy of Engineering, and members of the Design and Technology Association. The task of the group was to develop a statement of design and technology that emphasized the underpinning body of knowledge without compromising the unique nature of the subject, in which what a pupil can do with this knowledge is valued to the same extent as simply "having" that knowledge. A key feature of this statement is a design and technology "toolbox" of four elements: technological tools, for fabricating things and making things work; designing tools, for understanding needs, wants, and opportunities and for generating and developing ideas; critiquing tools; and data-handling tools. In developing this toolbox, the E4E has been at pains to define the knowledge necessary to understand and use the tools appropriately in response to a wide range of designing and making opportunities. Underpinning the use of the toolbox is the idea that successful design and technology requires action and reflection to work together in a concerted way, and without acquiring the necessary knowledge, neither activity will be successful and the overall endeavor will fail. It remains to be seen whether the minister for education will be convinced by these arguments.

The Design Council, which now incorporates the Commission for Architecture and the Built Environment, is another powerful stakeholder in design and technology education. It receives grants from the Department for Business, Innovation, and Skills and the Department for Communities and Local Government and has as a main aim to promote design and architecture for the public good. Recently it published the document "New Fundamental Principle for Design and Technology," which was endorsed by a wide range of bodies concerned with design, including the Council for Higher Education in Art and Design, the Creative Industries Council Skillset Skills Group, and the Royal Society for the Encouragement of Arts, Manufactures, and Commerce. The principles were organized into six groups: (1) to develop a design-literate society, (2) to build design and technology capability in its own right to act as a bridge between arts, science, and business, (3) to place human-centered design approaches, methodologies, and processes at the heart of learning, (4) to focus on technical skills that relate to design processes in three-dimensional, digital, and visual communication of ideas, (5) to embed design and technology within an academic and cultural framework, and (6) to forge strong links with industry and the cultural sector to inspire future designers, engineers, technologists, and manufacturers and introduce cutting-edge practices into the classroom. Again, the proposal of this new framework for the subject can be seen as an attempt to convince the minister for education that design and technology should be considered as worth including in the National Curriculum.

It would be of particular benefit to design and technology if these two powerful stakeholders, The Design Council and E4E, could be seen to agree on the substance of the subject with regard to the school curriculum in a way that enhanced the roles of both reflection and action. Inspection of the details in the proposals of the E4E and the Design Council shows that most of the Design Council's requirements for design and technology could be met by appropriate application of the E4E toolbox.

A recent development that has, as yet, had little impact on practice is the idea of open starting points for design and make assignments. This is part of the Design and Technology Association's modernization program and initially concerned the designing and making of electronic products. Six starting points were chosen on the grounds that they could lead to pupils' designing and making electronic products of varying complexity depending on the sophistication with which the pupils responded. Hence the starting points are not age or key stage specific. The six starting points identified were (1) playtime, (2) keeping in touch, (3) keeping secure, (4) staying safe, (5) thinking machines, and (6) other worlds. There are, of course, many other possible and valid starting points, but for the purposes of this exercise, this number was felt to be sufficient and provided a sufficient variety to be of interest and use to both teachers and pupils. On the Design and Technology Association website http://www.ectcurriculum.org, the starting points are presented as visual brainstorms, allowing the teacher and the class to explore the starting point for a wide range of possible briefs. These open starting points provide the opportunity to give pupils a voice as to what sort of product they want to design and make. The exact nature of the products designed and made will depend on the age and previous experience of the pupils and the resources available in the school, but giving the pupils a choice will provide ownership and is likely to increase their motivation. As pupils become used to such an approach, it is possible that they will not only be able to make choices as to the nature of the product they design and make but also have some autonomy in the way they go about this process. Reflection about the starting point and the possibilities it affords are clearly an integral part of the process leading directly to the pupils taking action in the manufacture of their design proposals.

DISCUSSION

The report *Assessment of Performance in Design and Technology* (Kimbell et al., 1991) acknowledged the importance of both reflection and action in pursuing design and technology activities. It must be remembered that at the time the research was carried out, a National Curriculum did not yet exist in England, hence none of the pupils had been taught design and technology as a school subject. The report discusses at some length the reflective-active balance that occurs in the responses of different pupil groups under different conditions in response to different tasks. One of the interesting findings was that for girls (who are generally seen as more reflective than boys) who had experienced a craft design and technology course (the subject most similar to design and technology), their ability to be reflective "enhances their active procedural capability." Even in these early days, the importance of reflection-action interaction was seen as significant. The tasks used to explore performance in design and technology were of necessity truncated, and it can be argued that to some extent they lack the authenticity achieved through extended design and make activities. Richard Kimbell (2004) has been critical of the way assessment practice in design and technology has developed since the publication of the APU report. He sees most current assessment practice as "widely regarded as having become formulaic, routinised and predictable" (Kimbell, 2004, p. 103) He takes this criticism further, stating, "It has become increasingly evident over the last few years that a number of pressures have combined to reduce learners' innovative performance at GCSE in design and technology. 'Playing safe' with highly teacher managed projects has been seen to be the formula for schools guaranteed A-C pass rate" (Kimbell, 2006, p. 18). In response to the concerns he raised, there has been renewed interest in assessment, and this has led to the Innovating Assessment research project (Golsmiths, 2009), which has produced structured testing materials to be used over a three-hour time block under examination conditions. This is a huge step forward compared to the stranglehold that the coursework portfolio has over pupil assessment in design and technology. Despite the undoubted achievement of this research project, there have to be reservations about a one-size fits all timed test pupils as the best way for pupils to reveal their capability in design and technology, particularly if consideration is to be given to the interplay of reflection and action and the time that this takes. Barlex (2007) has argued for a different approach in which the nature of the design decisions made by pupils tackling authentic tasks is the focus. He envisages a pupil making a series of "What if I did this" moves (Schön, 1987) as he or she considers possible decisions about a feature and its effects on decisions made or yet to be made about other features. This interconnectedness reflects a constructivist reflection-in-action paradigm for the pupil considering the process of designing as a reflective conversation with the situation (Dorst & Dijkhuis, 1995). Yet the utilization of a "What if I did this" strategy is more than a mere ad hoc tool to cope with the complexity of designing. Its repeated use increases the designer's understanding of the issues, thereby informing, guiding, and stimulating further designing both within and outside of the given design situation (Schön & Wiggins, 1992).

This view of pupil activity in design and technology has at its heart the interaction between reflection and action. It is worth considering this in the light of Lave and Wenger's (1991) idea of learning through participation in a community of practice. Those new to the community learn by taking part in the activities of the community. It is tempting to see pupils in a design and technology classroom as being members of a community of practice, under the guidance of the teacher, in which they tackle tasks that are personally authentic (in that the pupils are designing and sometimes making items that they feel are important) but also culturally authentic (in that they reflect to some considerable extent the professional practice of designers). A significant part of the learning that would take place as pupils develop the ability to respond fluently to such tasks would be to be both reflective and active in ways that enabled these behaviors to inform each other.

CONCLUSION

Some might argue that reflection and action are fundamental to design and technology and so intertwined that it is not possible for pupils to learn the subject without being both reflective and active in almost equal measure. However, this need not be the case. It is possible for pupils to have a skewed active-reflective balance such that their actions may be on the one hand too hasty and poorly thought through or on the other hand overconsidered so that the actions are insufficiently definitive. To some extent this can be overcome by supportive teacher guidance. A more significant problem is the availability to teachers of pedagogy that supports the development of reflection and action in a concerted way. Without pedagogy that deliberately sets out to achieve this, the roles of action and reflection become left to chance and pupils are unlikely to develop these essential twin attributes. The pedagogy described here provides teachers with the means to achieve this. At a time of uncertainty in England with regard to the future status and nature of design and technology in the curriculum, it is important that any new manifestations of the subject pay due regard to the inclusion of both action and reflection as key components.

REFERENCES

Banks, F., Barlex, D., Jarvinen, E., O'Sullivan, G., Owen-Jackson, G., &Rutland, M. (2004). DEPTH—developing professional thinking for technology teachers: An international study. *International Journal of Technology and Design Education, 14*(2), 141–157.

Barlex, D. (1995a). *Nuffield Design and Technology study guide.* London, UK: Longman.

Barlex, D. (1995b). *Nuffield Design and Technology resource task file.* London. England: Longman.

Barlex, D. (1995c). *Nuffield Design and Technology teacher guide.* London, UK: Longman.

Barlex, D. (1998). Design and technology: The Nuffield perspective in England and Wales. *International Journal of Technology and Design Education, 8,* 139–150.

Barlex, D. (2000). *Nuffield Design and Technology teacher guide.* London, UK: Longman.

Barlex, D. (2001). Young Foresight. *New Media in Technology Education, PATT11 (Pupils Attitudes Towards Technology) Conference Proceedings, 2001,* 31–33. Eindhoven, The Netherlands: University of Eindhoven.

Barlex, D. (2003). Considering the impact of design and technology on society—the experience of the Young Foresight Project. *The Place of Design and Technology in the Curriculum, PATT13 (Pupils Attitudes Towards Technology) Conference Proceedings 2003,* 140–144. Glasgow: University of Glasgow.

Barlex, D. (2004). Creativity in school technology education: A chorus of voices. *Learning for Innovation in Technology Education, 3rd Biennial International Conference on Technology Education Research Proceedings, 2004,* 24–37. Melbourne: Griffith University.

Barlex, D. (2006). Change of place. Retrieved from http://www.interactive.bis.gov.uk/foresight/TeachersGuide/Index.html.

Barlex, D. (2007). Assessing capability in design and technology: The case for a minimally invasive approach. *Design and Technology Education: An International Journal, 12*(2), 9–56.

Barlex, D., & Trebell, D. (2008). Design-without-make: Challenging the conventional approach to teaching and learning in a design and technology classroom. *International Journal of Technology and Design Education, 18,* 119–138.

Barlex, D., & Welch, M. (2001). Educational research and curriculum development: The case for synergy. *Journal of Design and Technology Education, 6*(1), 29–39.

Bronowski, J. (1973). *The ascent of man.* London, UK: British Broadcasting Corporation.

Department for Education. (2011). *The Framework for the National Curriculum. A report by the Expert Panel for the National Curriculum review.* London, UK: Department for Education.

Department for Education and Science and Welsh Office. (1988). *National Curriculum Design and Technology Working Group INTERIM REPORT*. London, UK: DES and Welsh Office.

Dorst, K., & Dijkhuis, J. (1995). Comparing paradigms for describing design activity. *Design Studies, 16*, 261–274.

Givens, N., &Barlex, D. (2001). The role of published materials in curriculum development and implementation for secondary school design and technology in England and Wales. *International Journal of Technology and Design Education, 11*(2), 137–161.

Goldsmiths. (2009). E-scape portfolio assessment. Phase 3 report. Commissioned by the British Educational Communications and Technology Agency (Becta). London, UK: University of London. Retrieved March 20, 2014, from http://www.gold.ac.uk/media/e-scape_phase3_report.pdf.

Kimbell, R. (2004). Assessment in design and technology education for the Department of Education and Skills. In *Learning for Innovation in Technology Education Conference* (Vol. 2, pp. 99–112). Brisbane, Australia: Griffith University.

Kimbell, R. (2006). Innovative performance and virtual portfolios—a tale of two projects. *Design and Technology Education; An International Journal, 11*(1), 18–30.

Kimbell, R., Stables, K., Wheeler, T., Wosniak, A., & Kelly, V. (1991). *The assessment of performance in design and technology*. London, UK: School Examinations and Assessment Council.

Lave, J., & Wenger, E. (1991). *Situated learning: Legitimate peripheral participation*. Cambridge, UK: Cambridge University Press.

Murphy, P., Lunn, S., Davidson, M., & Issitt, J. (2000). Young Foresight phase 1 evaluation report. Bucks, UK: Open University.

Murphy, P., Lunn, S., Davidson, M., & Issitt, J. (2001). Young Foresight summary evaluation report. Bucks, UK: Open University.

Murphy, P., & Hennessy, S. (2001). Realizing the potential—and lost opportunities—for peer collaboration in a D&T setting. *International Journal of Technology and Design Education, 11*, 203–237.

Ofsted. (2011). *Meeting technological challenges? Design and technology in schools 2007–10*. London, UK: Crown Copyright.

Schön, D. A. (1987). *Educating the reflective practitioner: Toward a new design for teaching and learning in the professions*. San Francisco, CA: Jossey-Bass.

Schön, D. A., & Wiggins, G. (1992). Kinds of seeing and their functions in designing. *Design Studies, 13*, 135–156.

CHAPTER 9

FROM CYBEREDUCATION TO CYBERACTIVISM

CAN CYBERLITERACY TRANSFORM THE PUBLIC SPHERE?

Andoni Alonso

Technology is commonly imagined as liberating. This is especially true when thinking about information and computer/communications technology (ICT) and the construction of cyberspace, a realm seemingly free of physical constraints. (Debates about the precise meaning of "ICT," and whether the C stands for "computer" or "communication", suggest a preference for "cyber-" constructions.) The situation is hypertrue when imagining cybereducation—since public education is the foundation of any strong public-sphere cyberactivism. The reality, however, is more problematic— especially when considering the central role of cyberscience. Liberation is never the automatic result of technological change.

CYBEREDUCATION

From the beginning, information and computer technologies were projected to be powerful means for opening up higher education and lowering costs (Feenberg, 2001). New electronic devices promised an educational utopia: the possibility of an unfettered dissemination of knowledge and literacy. The idea is that access and costs are the primary barriers to making the university democratically available to everyone and that the former can be made virtually present everywhere while simultaneously reducing the latter to virtually nothing. Additionally, much has been written about how the information society and economy depend on knowledge as a key resource (Castells, 2001). Online education and cyberliteracy came to be thought of as the foundations for contemporary society (Echeverría, 1999 and 2003).

Of course, cyberliteracy is not the same as literacy in the traditional sense of reading, writing, and critically interpreting texts. The new techno- or cyberliteracy involves document accessing; word processing; emailing; database, spreadsheet, and PowerPoint slide creating; and more. Some argue that even programming is required—as if driving a car required knowing how an internal combustion engine works. Certainly web page design has become a commonly assumed skill, as well as knowing how to protect against malware infections. Even more basic is the fact that people must be able to access computer hardware. Challenges associated with the digital divide between the hardware rich and hardware poor have cast a skeptical shadow over the dream of cybereducation (Norris, 2001).

The digital dream has nevertheless strongly influenced the educational system in Spain, where the Universidad Nacional de Educación a Distancia (UNED, or National University of Distance Education) was founded in Madrid in 1972 by educational technocrats who were ascendant during the last years of the Franco regime. UNED now has the largest enrollment of any university in Europe, with more than 250,000 students and outreach centers in 13 countries stretching into the Americas and Africa. The Universitat Oberta de Catalunya (UOC, or Open University of Catalonia), established in Barcelona in 1994 as the university of the knowledge society is another institution attempting to utilize ICT to transform higher education; Manuel Castells has lent his prestige as a UOC professor.

Yet mere quantitative increases in student numbers obscure as well as disclose. History, for instance, reveals the repetitive character of the utopian discourse about educational technologies—and how much current efforts echo previous fantasies with regard to earlier communications media (see Waks 1995). Every new electronic communications invention, from radio and motion pictures to television and the Internet, has been invested with the same promise: the technology will expand higher education as never before, increasing the productivity of teachers while making their work globally accessible and inexpensive to students. Democratic intelligence will blossom forth on the planet. At the same time, no more than a brief look at course offerings of the online educational programs—many of which are actually offered in the United States by private, profit-making corporations—discloses a shift away from the humanities and the social sciences, which are the foundation of democratic intelligence, toward technical and vocational training. Cybereducation commonly focuses on developing skills related to the processes of cybereducation—computer programming, web design, database management, and related activities—instead of cultivating the critical skills associated with traditional literacy along with reflection on the goals, goods, and aims of life.

Education itself can be conceived in two quite different ways: as a fundamental right or as a commercial product. According to the *Universal Declaration of Human Rights* (1948), Article 26:

(1) Everyone has the right to education . . . Technical and professional education shall be made generally available and higher education shall be equally accessible to all on the basis of merit.

(2) Education shall be directed to the full development of the human personality and to the strengthening of respect for human rights and fundamental freedoms. It shall promote understanding, tolerance and friendship among all nations, racial or religious groups, and shall further the activities of the United Nations for the maintenance of peace.

It is not clear that cybereducation goes beyond extending technical and professional education—often for the commercial benefit of cybereducation providers.

The fundamental rights view also expresses a traditional public goods concept. According to John Dewey, for instance, "education is a necessity of life" and crucial for democracy: "A society which makes provision for participation in its good of all its members on equal terms and which secures flexible readjustment of its institutions through interaction of the different forms of associated life is in so far democratic. Such a society must have a type of education which gives individuals a personal interest in social relationships and control, and the habits of mind which secure social changes without introducing disorder" (Dewey, 1916, chapter 7, summary).From Dewey's perspective, education may also be described as a commons—and could be analyzed using Elinor Ostrom's (2000) theories of collective action, trust, and the cooperative management of common pool resources.

Any society contextualizes education; education is always part of a larger culture, and contemporary European and North American cultures commonly fail to conceive of themselves (European perhaps less than American) in common pool resource terms. In our turbo-capitalist societies, educational technology acquires a distinctive shape. In the United States, given its long-term capitalist ideological commitments, there has always been a small for-profit education sector, mostly for technical training. But since the 1990s, as a by-product of the neoliberal capitalist movement to deregulate commercial activities and privatize public services, the traditional not-for-profit institutions of higher education, both public and private, have increasingly been complemented with for-profit universities. In the 1970s, only 2 percent of US students were enrolled in for-profit higher education institutions; by 2008, this had increased to 8 percent (Geiger & Heller, 2011). Cybereducation has been a major contributor to this increase.

Cybernetworks may have been imagined as a means for creating universal access to traditional literacies. Internet use itself may have been thought sufficient to achieve utopian ideals (Negroponte, 1995). The Internet as a global repository for all human knowledge may make it theoretically possible for anyone to become an intelligent citizen or even a critical intellectual. Yet the realization of such possibilities on any mass scale is quite improbable (Maldonado, 1995).

Online packaged education has, in practice, often become a way to co-opt the free time of citizens in the acquisition of skills necessary for keeping up with technical change—the colonization of leisure—while outsourcing the work of professors and depersonalizing learning. Cybereducation cannot substitute for face-to-face discussion, although social media can sometimes enhance it. It is difficult to understand the excitement in the humanities about Massive Online Open Courses (MOOCs), since these are really no more than canned lectures broadcast to thousands. What is actually happening is revealed by the excitement of venture capitalists who are willing to invest in MOOCs as potentially new sources of revenue and profit.

The free and open source software (FOSS) movement and Wikipedia point toward alternative models for cybereducation. Software is as important as hardware for the construction of cyberspace, but business models for software production emerging from corporations conflicted with the more spontaneous sharing characteristic of premonetized software development. A nonmonetizing model of software creation was at the center of the FOSS movement (Stallman, 1999). At the time, most people thought it impossible to develop an alternative to the UNIX operating system, which AT&T privatized in 1982 as part of its agreement with the US Department of Justice to break up the company and spin off the Bell Labs research division. Bell Labs then had to make money and become self-supporting and saw UNIX licensing as part of its new business plan. But the computer ethic and ideal of collaborative sharing, of having a common task to which anyone who was willing might contribute, was so strong that it attracted more than a hundred thousand individual programmers and actually worked (Himannen, 2001).

The first e-learning platforms, such as PLATO (Programmed Logic for Automatic Teaching Operations) at the University of Illinois, were constructed on the FOSS model. (When PLATO was commercialized by Control Data Corporation, it failed.) Moodle (Modular Object-Oriented Dynamic Learning Environment), currently the most popular learning management system, is also a FOSS product, first released in 2002 by graduate computer science student Martin Dougiamas in western Australia. As of mid-2013, its user base was more than 80,000 sites, while that of its privatized, commercial competitor Blackboard was approximately 37,000 sites. Growing up in a small desert settlement, Dougiamas received his primary school education via shortwave radio distance education supplemented by regular airplane book drops; his desire was to bring this kind of noncommercialized learning to the Internet, as well as learning management systems. Although the noncommercial model of Moodle is currently in ascendance, it has to compete with the commercialization of Blackboard.

Another model for a public good, common pool cybereducation, can be found in Wikipedia. The wiki software for collaborative content creation was developed by extreme computing programmer Ward Cunningham beginning in the mid-1990s. In 2001, philosopher Larry Sanger and Internet entrepreneur Jimmy Wales conspired to use wiki software as the foundation for a new kind of online reference work that would draw on the knowledge base of anyone, anywhere who wanted to contribute. As of mid-2013, according to Wikipedia itself (which has repeatedly been shown to be at least as reliable as the *Encyclopedia Britannica*), Wikipedia has "30 million articles in 287 languages, including over 4.3 million in the English Wikipedia, [which] are written collaboratively by volunteers around the world. Almost all of its articles can be edited by anyone having access to the site. It is the largest and most popular general reference work on the Internet, ranking seventh globally among all websites on Alexa, and having an estimated 365 million readers worldwide."

Reading Wikipedia—and even more, contributing to it—constitutes engagement with a common pool, public good, educational experience. Wikipedia is the most visible manifestation of education or knowledge acquisition and production as a freely available right. Again, however, Wikipedia exists in tension with commercial educational content, software, and service providers such as Cengage Learning and Pearson, the former of which markets proprietary reference works such as the *Encyclopedia of Bioethics* and the *Encyclopedia of Science, Technology, and Ethics*.

CYBERSCIENCE

The two conceptions of education as public good and as commercial product have close analogues in that distinctly modern form of knowledge acquisition and production known as *science*—including cyberscience, that is, science transformed by ICTs. From its earliest forms, modern natural science was inherently technological. Its theory was oriented toward production (for the "conquest of nature," in the words of Francis Bacon); it was dependent on instrumentalization (Galileo Galilei's telescope and Antonie van Leeuwenhoek's microscope); and it even created new sciences of particular technologies in order to improve them (William Rankine's thermodynamics of James Watts's steam engines). As this interrelationship has become thematized, science is increasingly described as technoscience—and the philosophy of technology supersedes the philosophy of science as an effort to understand the contemporary world.

In one sense, cyberscience is simply a new form of technoscience. To the methods of doing science by mathematical theory construction (deduction) and by empirical observation and experimentation (induction), computers have added a third form: by simulation. Simulations create virtual or cyberworlds that can be used to give mathematical theories dynamic graphical representations in which virtual phenomena testing can take place. Global climate models (GCMs) exemplify cyberscience in this sense. Another form of cyberscience in this sense attempts to reform science education to make it more effective and efficient.

A second type of cyberscience builds on information and communications technologies. Large-scale collaborations, even transnational teams, cooperate on small parts of big science projects such as sending space probes to Mars or mapping and sequencing the human genome. This is cyberscience enabled by prototypes of social media. Social media cyberscience is also transforming the peer review process and scientific publishing (open, interactive peer review and open access journals). The result is sometimes just called open science or Science 2.0.

Light can also be thrown on what is taking place in science by Yochai Benkler's (2006) analysis of transformations operative in the cybernetworked economy. From the past are efforts to defend intellectual property rights with patents, copyright laws, and trade secret agreements. Stimulated by the emergence of cyberspace are efforts to practice the free sharing of knowledge in a kind of digital gift economy. The former continues to envision a market logic based on scarcity; the latter sees a market emerging based on digital affluence. It is within the context of potentials for digital affluence that both types of cyberscience arise.

The first type of cyberscience characteristically remains more professionalized. Although not privatized in the commercial sense, it is nevertheless privatized in another sense: It takes place in a professional realm and gives science as a regionalized public sphere an increasingly cyberistic cast. The civil society of scientists that previously existed in the classroom, the laboratory, and scientific conferences is projected into cyberspace. Simulations can even be thought of as cyberized field work—field work being another locus of scientific civil society distinctive of environmental scientists. Cyberscientific efforts in science education reform, using computer models and clickers or other more advanced (and invasive) devices to control student learning, are even more obviously based in the professional scientific community.

The second type of cyberscience opens the door to nonscientist participation in science and what has also been called "citizen science" and public participation in technological decision making. Citizen science and technology can take top-down and bottom-up forms. One top-down example of citizen science is associated with efforts by the Laboratory of Ornithology at Cornell University to enroll amateur scientists in the collection of data that could then be analyzed by professional scientists; indeed, some evidence points toward Rick Bonney, working at the Cornell Lab in the 1980s, as the person who coined the term "citizen science." Yet the practice itself is much older; see, for instance, the Christmas Bird Count of the American Audubon Society, which goes back to 1900. Now such citizen-science projects utilize smartphones and related electronic devices for higher-quality data collection and more rapid communication. Citizen science projects dependent on ICTs include SETI@Home and the search for extraterrestrial life, volunteers working for NASA to classify astronomical objects, and the monitoring of flower pollinating animals—all can be accessed through the *Scientific American* "Citizen Science" web site (http://www.scientificamerican.com/citizen-science).

Bottom-up citizen science is more activistic and can even take the form of do-it-yourself science by nonprofessional scientists. Again, there are pre-cyberspace anticipations. The AIDS activist movement of the 1980s, which demanded shifts in the priorities and practices of HIV-related research, became a paradigmatic case in which people with nonscientific vested interests in a research program forced scientists and policy makers to consider their own perspectives (Epstein, 2004). Since then, cyberspace and social media have definitely enhanced opportunities for citizen-initiated public-sphere technoscientific policy making.

Cyberactivism

Bottom-up citizen science is a form of cyberactivism. As Wikipedia describes it, cyberactivism (broadly construed as synonymous with Internet activism, online activism, digital campaigning, digital activism, online organizing, electronic advocacy, e-campaigning, and e-activism) constitutes "the use of electronic communication technologies such as social media, especially Twitter and Facebook, YouTube, e-mail, and podcasts for various forms of activism to enable faster communications by citizen movements and the delivery of local information to a large audience." This is an unfortunately thin description that ignores its more substantive features. In the present instance, the focus is on cyberscience activism, an aspect of cyberactivism that is seldom otherwise discussed, as a necessity emerging from efforts to address substantive public-sphere problems.

Prior to the creation of cyberspace, bottom-up activism was prefigured by top-down science activism. Nuclear scientists and engineers who created the Federation of Atomic Scientists, the *Bulletin of Atomic Scientists*, and later the international Pugwash movement to lobby against nuclear weapons testing and proliferation provide one example. Another was biologist Rachel Carson, who sought to popularize her scientific knowledge to alert the public of the dangers of chemical pollution so that those affected could take political action to limit pesticide contamination of the environment. In cyberspace it has become possible—indeed, necessary—to complement such top-down noblesse oblige scientific activism with bottom-up public activism that both uses and provides guidance for technoscience.

Cyberactivism in this new sense also depends on cyberliteracy that goes beyond technical tool learning for more effective performance. Laura Gurak, for instance, argues that "what we really need to understand is not just how to use the technology but how to live with it, participate in it, and take control of it" (Gurak, 2003 p. 11). Cyberliteracy must help people develop critical perspectives on such issues as the speed, reach, and interactivity of the Internet; privacy and digital rights; and how to distinguish genuine information from fabrication, falsification, and plagiarism (to reference the standard forms of scientific misconduct).

Bottom-up, citizen-initiated cyberscience activism is grounded in efforts to exercise a right closely associated with cybereducation and cyberscience:

the right to know. This is a right not limited to scientists and their acquisition or production of knowledge; neither is it equivalent to scientists' contestable claims concerning a right to do any research they want—even when enhanced by the top-down management of citizen support. Cyberactivism in the strongest sense rests on peoples' right to participate in and influence those decisions and activities that affect them (see Mitcham, 1997).

This right to participate also entails a right to knowledge that makes such participation intelligent and effective. More generally, according to the *International Covenant on Civil and Political Rights* (adopted 1966, in force as of 1976), Article 19, "Everyone shall have the right to freedom of expression; this right shall include freedom to seek, receive and impart information and ideas of all kinds, regardless of frontiers, either orally, in writing or in print, in the form of art, or through any other media of his choice."

Although this right to know bears primarily on the political information necessary to enable intelligent democratic behavior, given that civil society is increasingly a locus of scientific and technological action, it must be extended to scientific and technological knowledge.

In most advanced technoscientific countries, a "right to know" related to technoscientific knowledge takes two forms: one related to the workplace, another to the environment. It is now a generally accepted legal principle that workers have a right to know the chemicals to which their jobs may expose them and that citizens have a right to know any hazards to which their communities may be exposed. Both rights to know were initially promoted by top-down noblesse oblige scientific activism.

Enacting these rights to scientific knowledge is nevertheless complicated by multiple factors. One major issue is the extent to which scientific research is now influenced by private money. Since the mid-1980s, the proportion of scientific research funded by the private rather than the public sector has significantly increased. In the academic world, young career researchers have increasing difficulty securing research support. One empirical study of early and midcareer scientists found that 15 percent reported modifying the design, methodology, or results of their work as a result of pressure from a funding source (Martinson et al., 2005). More dramatically, journalists have described the rise of "junk science" in the courtroom, as scientists are hired by attorneys to testify for their clients (Rampton & Stauber, 2001). Indeed, there is a growing general distrust of scientific experts and risk managers (Collins &Evans, 2009).

The definition of junk science is contested in ways that further highlight the problem. Although the term initially emerged in the courtroom to disparage paid expert witnesses defending corporate interests, it has come to be applied by corporate interests to any science that might threaten corporate behavior. Scientists defending tobacco companies have been accused of purveying junk science, but so have members of the Intergovernmental Panel on Climate Change.

The need for scientific testimony points toward a second difficulty in securing rights to scientific knowledge. In traditional legal cases, expert witnesses are seldom necessary. If someone harmed another's person or damaged property, common sense could usually determine who was responsible and at what level. Over the course of the twentieth century, however, personal harm and property damage by technoscientific activity became increasingly difficult to determine. When injuries are caused by radiation from nuclear weapons development or tests or the release of toxic chemicals from industry or involve complex dose-response relationships over extended periods of time (as with asbestos-related mesothelioma), then science is necessarily required to help determine legal liabilities. Common sense is often incapable of dealing with the complex consequences of technoscientific activity.

At the same time, the increasing role played by technoscience in public-sphere discussions of economic affairs, health care policy, and national defense makes citizens increasingly in need of scientific knowledge. Concerned citizens have to organize to understand what is going on, how they can change things, and how they can have a better life than markets and technoscientific systems alone are able to construct. Questions go beyond any simple enjoyment of the benefits of science and technology; citizens also need to protect themselves and to improve decisions. Cyberactivism becomes not just a possibility but a duty.

CONCLUSION

Originally associated with visions of higher education as a common good to be made more universally available through gift-economy collaborative production, cybereducation is threatened by the colonizing interests of private, profit-generating commerce. Analogous tensions exist in cyberscience between a top-down utilization of ICT to enlist nonscientists in scientific knowledge production and bottom-up efforts by cyberactivists to make science more responsive to public needs. In a public sphere increasingly influenced by technoscience, it is crucial for cybercitizen activists to be able to draw on science to enhance the intelligence of their participatory engagement with science policy and technological decision making. Cyberliteracy—including awareness of the tensions and complexities outlined in this chapter—can help enhance the potential for positive transformations of the public sphere. Such transformations nevertheless remain always fragile and contingent.

ACKNOWLEDGMENT

This article has grown out of extended exchanges with Carl Mitcham and has benefited from his English-language editing.

NOTE

All quotations from Wikipedia were accessed in September 2013.

REFERENCES

Benkler, Yochai. (2006). *The wealth of networks: How social production transforms markets and freedom.* New Haven, CT: Yale University Press.

Castells, Manuel. (2001). *La galaxia Internet*. Barcelona, Spain: Plaza y Janés.

Collins, Harry, & Evans, Robert. (2009). *Rethinking expertise*. Chicago, IL: University of Chicago Press.

Dewey, John. (1916). *Democracy and education*. New York, NY: Macmillan.

Echeverría, Javier. (1999). *Los señores del aire: Telépolis y el Tercer Entorno*. Barcelona, Spain: Destino.

Echeverría, Javier. (2003). Tecnociencias de la información y participación ciudadana. *Isegoría, 28*, 73–92.

Epstein, Steven. (2004). *Impure science: AIDS, activism, and the politics of knowledge*. Berkeley, CA: University of California Press.

Feenberg, Andrew. (2001). "La educación y las opciones de la modernidad." In Andoni Alonso & P. Blanco (Eds.), *Pensamiento digital. Humanidades y tecnologías de la información* (pp. 115–133). Badajoz, Spain: Junta de Extremadura.

Geiger, Roger L., & Heller, Donald E. (2011, January). Financial trends in higher education: The United States. *Peking University Education Review, 9*(1), 15–32, 187–188.

Gurak, Laura J. (2003). *Cyberliteracy: Navigating the Internet with awareness*. New Haven, CT: Yale University Press.

Himannen, Pekka. (2001). *The hacker ethic and the spirit of the information age*. New York, NY: Vintage.

Maldonado, Tomás. (1995). *Che cos'è un intellettuale? Avventure e disavventure di un ruolo*. Milan, Italy: Feltrinelli.

Martinson, Brian C., Anderson, Melissa S., & de Vries, Raymond. (2005, June 9). Scientists behaving badly. *Nature, 435*, 737–738.

Mitcham, Carl. (1997, Spring). Justifying public participation in technical decision making. *IEEE Technology and Society Magazine, 16*(1), 40–46.

Negroponte, Nicholas. (1995). *Being digital*. New York, NY: Random House.

Norris, Pippa. (2001). *Digital divide: Civic engagement, information poverty, and the Internet worldwide*. Cambridge, UK: Cambridge University Press.

Rampton, Sheldon, & Stauber, John. (2001). *Trust us, we're the experts! How industry manipulates science and gambles with your future*. New York, NY: Tarcher/Putnam.

Ostrom, E. (2000). Collective action and the evolution of social norms. *The Journal of Economic Perspectives*, 137–158.

Stallman, Richard. (1999). The GNU operating system and the free software movement. In Chris DiBona, Sam Ockman, & Mark Stone (Eds.), *Open Sources: Voices from the Revolution* (pp. 1–21). Sebastopol, CA: O'Reilly.

Waks, Leonard. (1995). *Technology's school: The challenge to philosophy*. Research in philosophy and technology, supplement 3. Greenwich, CT: JAI Press, 1995.

TECHNOLOGICAL LITERACY
IN THE WORKPLACE

CHAPTER 10

SITUATING TECHNOLOGICAL LITERACY IN THE WORKPLACE

Jamie Wallace and Cathrine Hasse

The developing discourse centered around the definition of technological literacy has been taken up from many quarters and between conflicting perspectives (Kahn & Kellner, 2006; Keirl, 2006). While the array of different technologies and applications that might be considered gives rise to differing disciplinary viewpoints (Liddament, 1994), there remains a lack of an adequate framework from which to view technology and its use within a particular context of practice. Seeing this as central to the development of professional disciplines immediately places technology literacy not simply as something useful for ensuring that certain technologically mediated tasks can be adequately satisfied but rather, because of technology's pervasive quality, as something that encompasses the nature of working life itself.

Arguments as to what technological literacy should be have primarily turned toward the educational system (e.g., Garmire &Pearson, 2006). Beyond this, there has been considerable consideration of how the results of education can enable a meaningful engagement with the progressively technological world (see Ingerman & Collier-Reed, 2011). There has, however, been little direct, empirically informed understanding of technologies' consequences for the various wide-ranging concerns of working practitioners. In this sense, notions of technological literacy have remained abstracted from situated everyday working experiences beyond the scope of the isolated operation of technologies themselves.

Considering technological literacy as something being realized within work situations, among other things, provokes the question of how it can be identified and studied. As no single body of attributes or characteristics is as yet able to successfully point to what it means to be technologically literate in everyday practices, then, if the term is to move beyond a merely aspirational endeavor, the narratives of those engaging purposely with technology can be seen as a way of understanding how it plays a part in situated practice. This approach relates to the ways workers express their own understandings of technology as they engage with it in their everyday practices.

The understanding of technology's position within the workplace has been studied widely from different quarters, such as those of STS (science and technology studies) and CSCW (computer-supported cooperative work). Although these unearth much that can inform us of the situated role of technology, there has been little in this direction specifically related to technological literacy within professional work contexts. This is clearly demonstrated in the fields of both nursing and teaching. There are many studies that tackle issues related to the influence of "electronic patient journals" in hospitals or consider, for example, the didactic improvements to "interactive whiteboards" within schools, but there is little focus on the broader needs of practitioners to integrate and align these with the wider concerns of their everyday lives. In this respect, we consider technologies not as isolated tools freely adopted and discarded at will but rather as materials and systems of understanding variously embedded within working life and mediating the progressive reconfiguration of procedures, processes, and structures. Technologies are therefore continually being realized within work situations, raising important questions as to how they are best understood in relation to their multiple consequences.

In the following, we draw on empirical work aimed specifically at considering the need to develop technological literacy among the professions of nursing and teaching (Wallace, 2012; Tafdrup & Hasse, 2012). Central to this is the study of how technology has been reconfigured within the everyday workplaces of Danish hospitals and schools coincident with the changing practices they influence. Technological literacy is understood not as an ability or competence derived solely through prior educational means but as something continuously developed in the workplace through practice-based learning in the face of the constant mutual reconfigurations of technology and practice. It becomes an important aspect of professional expertise to handle and negotiate these reconfigurations without losing sight of the motive for the practice itself.

What we propose here is a shift of focus from technology as something defined in disciplinary, educational, corporate, organizational, design, or even social terms. Alternatively, the attempt is to grasp technology as continually playing shifting and emergent roles within ongoing and recurrent interactions across a sphere of professional workplace influence. In this respect, we align with those scholars who see technological literacy as a way of acting and, as expressed by Ingerman and Collier-Reed (2011 p. 141), as "something that is realised in particular settings and situations-over and over again." Technological literacy doesn't result from an understanding of what technology is, whether from a historical viewpoint (Feenberg, 2006) or that of technological knowledge (De Vries, 2006), or that it is socially constituted. Rather, it exists in relation to the unfolding consequences of processes and ways of thinking and organizing mutually constituted between social and technological worlds.

This temporal aspect underlines the learning that springs from having to reapproach the position of "technologies-in-practice" (Dourish, 2006, p. 6). During the course of practice-based learning, working life emerges as a collective learning space where people find ways to learn from each other in the face of technologies influencing particular work-related situations. What becomes of interest here is whether there are overarching situations to which we can address the focus of "in-practice" technological literacy. In other words, are there central types of reconfigurations that would allow practitioners to pinpoint particular needs or characteristics of their mutual learning?

Difficulty immediately arises here in attempting to identify the general from the particular. Understanding technological literacy in general terms removes the very dependence on the situatedness necessary for actually engaging with technology. Although we can refer to notions such as "the ability of a person to use, manage, assess, and understand technology" (Duggar, 2001, p. 515), these remain detached from the in-practice engagements involving explicit and implicit forms of knowing that include deviating ways of experiencing and acting. Disciplinary boundaries are drawn between workers that give rise to "figured worlds" (Holland et al., 1998), and they inhabit different "ecologies of practice" populated with different technologies (Wallace, 2010). This leads to collaborative technological uses amid various forms of knowing and the multiple ontologies of working life. In order to make sense of technological literacy in these complex contexts, we look for the existence of patterns across situational encounters within which practitioners and technologies undergo transformations of becoming. These are recurring occurrences within everyday work experiences that can be identified as concurrent reconfigurations of technology and practice. Therefore, examples can shed light on the ways humans and technologies become distinctively coupled under certain workplace circumstances.

Yawson considers the principal ambition of technological literacy as providing people with the "tools to engage intelligently and conscientiously in the word around them" (Yawson, 2010, p. 5). What we want to set in focus here is not just something of the nature of these tools but how they apply differently as the world around them changes. Being neither static nor generalized, they place complications on how to develop meaningful ways of engaging. What may appear intelligent and conscientious at one juncture might seem very different at another. Exploring the distinctive impact these situated worlds have on ways of learning together with technologies provides a move toward situated approaches to technological literacy.

This type of practice-based learning stands in a close relationship with the concept of situated learning through which, as noted by Stephen Billett, occupational expertise develops in practice. It takes time and experience, as well as critical insight, to develop occupational skills, which are "shaped through particular episodes of experiences that comprise situated instances of practice" (Billett, 2009, p. 833). Learning here is situated as an integral part of generative social practice in the lived-in world (Lave & Wenger, 1991). It is in this situated activity that staff attempt to make sense of the technologies often imposed by politicians and managers and tied to dreams of increased efficiency and innovation.

This kind of response to change can be understood through theories of situated learning (e.g., Lave & Wenger, 1991; Brown & Duguid, 1991; Orr, 1996), as well as the acknowledgement of the discrepancy between explicit plans and situated actions (Suchman, 2007) that can be exposed following ethnographic studies. All these practice-based approaches underline the importance of ongoing situated changes and learning when work takes place. We see technological literacy as an integrated aspect of this process of change. Theories of practice-based learning, however, generally lack a perspective on how both technology and work life reconfigure each other in these learning processes—in other words, the understanding that accompanies the mutual reconfigurations and their dissipating consequences identified and contemplated through reflections in practice and on practice.

Michael sees this move toward the substantive as a part of technological literacy beyond that of technical knowledge and the mastery of technique (Michael, 2006). What becomes interesting, however, is how such things as, for example, "a commitment to discrimination, to improvement of the world, to criticism, creativity, and autonomy" (ibid, 56) are not separate from the understanding of the technologies but able to be conceived of as a mutual part of the technosocial landscape that unfolds with the changing forms and usage of technological artefacts, systems, and knowledges. Within the world of work and professional expertise, there are frameworks for acting that pose challenges to how we can understand the emergent and repetitive character of acting with technology that relates the intentions of technological use on the one hand to an understanding of the consequences on the other.

The fields of teaching and nursing offer two diverse examples of how broad changes to the technologies deployed within workplace settings lead to consequential adjustments of both working practices and the wider notions of the professions themselves. Technology can influence not simply practice but also ideas of what it is to be a practitioner (Barnard, 2006). As well as being socially dominant professional disciplines, they reflect a central dimension of technology's place within the workplace. Although increasingly influenced by technological use, both professions have as their underlying aspiration to fulfill what might be termed basic human pursuits and goals. What becomes challenging in this regard is the use of technology within a developing technological context without overtly or at least detrimentally influencing the attention to human aspects, be it from an operational viewpoint or related to the developing professional knowledge and skill sets. Significant to this is whether technologies are considered to have dehumanizing or disorienting effects detracting from the ultimate caring or learning aspiration of practice.

The working lives of teachers and nurses and the technologies they adopt exist within organizational and institutional contexts. They are subject not simply to the day-to-day concerns of the tasks to which they are set but to the wider workings of the organization. The use of technologies includes, in some measure, aspects beyond the functions for which they are deployed, such as economic, political, or ethical functions. Having a bearing on the responsibility of multiple stakeholders, technologies become boundary objects (Star, 1989; Carlile, 1997) whose influence spills over to forms of administration, accountability, communication, knowledge sharing, security, investment, and maintenance, as well as the very means of employment and its bearing on liabilities (Barnard, 2006). Given such complex interrelations and traditions, differing professions, institutions, and departments are subject to very different forms of organization and organizational culture. Considering, for example, the degree to which practices are subject to tight procedural control draws a divide between nursing and teaching. Nursing operates within strict guidelines and documented procedures. This isn't to say nurses aren't able to change the manner in which they operate and adopt technologies, but nursing has very different disciplinary circumstances than teaching. It is a profession that relies on what, on the face of it, we might call wide autonomy of practice. Therefore, we see differences in the approach to vocational and workplace learning and consequently how people learn to understand technologies "inpractice."

What the professions have in common is the pervasive influence of technological development, whether this is seen through the paradigm of innovation or not (Edgerton, 2007). Nurses find and update their patients' records through a specifically tailored computer system; they use electronic infusion drips, dynamic beds, electronic blood pressure measurement apparatuses, and small handheld computers. In schools, we see computer systems and the Internet connecting parents, pupils, and teachers; there are interactive whiteboards with increasingly new software features, copying machines, movie cameras, and most prominent of all, students' cell phones being included within teaching.

The progressive nature of technology isn't just visible in the changing styles adopted by designers and through the advanced capabilities of new technologies, but it becomes prominent in the narratives that practitioners give about their working lives. Given the opportunity to refer to all sorts of tools relevant to their work that might include books, pencils, and paper, it remained electronic artefacts and the very latest digital technologies that were considered as "technology." Asked to name the three most important technologies in their working lives, they pointed to things needing electrical power, such as computers, software systems, interactive whiteboards, or electronic defibrillators, while making associations with the practicalities of use (Tafdrup & Hasse, 2012). One nurse noted, "Well, I would answer 'the computer' because you use it all the time." In the words of a teacher, "Technologies? There is of course the computer, right? It is a working tool as well as a planning tool." Although different technologies are referred to in the different professions, it is more often than not the archetypical computer that is first to be referred to. On further reflection, it becomes particular programs and systems that are discussed or the use of specific artefacts overshadowing ideas of the generic computer.

Considering technological literacy as the mastery of computer skill, for example (see Waks, 2006), discounts not simply any contextual notion of the uses to which the computer is put but also the changing technological context of the artefact itself. It takes an artefact-centric approach in which the computer is considered a static artefact influenced by neither progressive innovative developments nor changing perceptions about its use, function, or wider consequence. In other words, we see a view of technology as removed from the significance it has to, for example, the ways people work, the time they use, how they relate to each other, how they perceive their working environment, and how they organize their daily routines, not to mention the demands for learning new competencies alongside fulfilling work obligations.

Aspects such as this show how professional judgments are multidimensional; people are able to see technologies as having multiple influences beyond those responsible for their capital investment. They develop in-practice relations through which they undergo a transition from the designed products to tools imbued with local sense and meaning. What we may generally call "the computer" can therefore demand different or multiple technological literacies (Kahn & Kellner, 2006) in different local situated contexts—even within the same profession.

In addition to the progressive or at least changing nature of technology, practitioners must learn to approach their influence at particular times as being uncertain. This may call for a marked reconfiguration of practice as the technology's response or consequence begins to deviate from what was expected. In these situations, complexity and uncertainty become different sides of the same coin as workers explore new opportunities to fulfill the unfolding demands of the situation. There are differences here between the familiar and the unfamiliar, between the imposed and the preferred (Wallace, 2012), or between the functioning and the broken. The locus here is the experience of engagement supported through cycles of learning that point to reasonable forms of reconfiguration.

For the worker unable to attempt his own repair, the notion of the "breakdown" becomes the severance of technology and practice. For those familiar enough with the technology's workings, a breakdown becomes another reconfiguration aimed at returning the technology to its working state. In some cases, a breakdown calls for a complete reevaluation of the intended procedure that can lead to the exercise of innovative solutions to ensure that the primary goal of practice can be maintained. If such things can be considered before the event of a breakdown, they might lead to the preparation of contingencies, of having a "plan B," or at least the awareness of having to "think on one's feet."

Both nurses and teachers refer to commonly experienced types of breakdowns (more than 80 percent mention breakdowns as a major problem for working life when asked about the biggest hurdle with using technologies; Tafdrup & Hasse, 2012), such as the failure of computer systems. Handling these becomes an integral part of their acting with technology and, as such, relies centrally on their situated technological literacy. Without appropriate action, breakdowns can influence multiple consequences. As a boundary object, a technology's failure is seen from different quarters in different ways. It might be a question of reduced efficiency, security, safety, loss of time and resources, or communication, or it might be seen as an opportunity to teach others how to act when failures occur. The resulting need for a fluency in navigating consequences involves learning "knowing how to know how" as it has been phrased by Anne Edwards (2010). Practitioners without this aspect of technological literacy risk ending up as passive users in the face of breakdowns, struggling to find alternatives to their suspended actions. For some, it might be an opportunity to draw management's attention to the need for improvement or to make a case for alternative technologies. For the practitioner, the breakdown of technology remains first and foremost a hindrance to the completion of the task for which the technology was being employed. The need for a reflective response places these situations at the forefront of the pragmatic ambitions of a situated technological literacy. It is no use repeatedly pressing the same keyboard button or computer screen command if there is no response! But does the answer lie in further engagements with the computer or in turning attention toward human activity?

This dilemma of attention between the human and the technological becomes another central theme within a situated technological literacy. For example, are the particular, rational demands of computer interfaces and the step-by-step logic of systemic procedures altering our ways of perceiving other nontechnological tasks? This may be difficult to say, but calling into question of the appropriate use of technological involvement must surely be relevant during all practices for which there are nontechnology alternatives. As expressed by several nurses, "There are times where clinically I think 'you can't always rely on technical equipment.'" Teachers, too, are increasingly aware of the consequences technologies are having on the ways they relate to their classes through, for example, the expectations pupils have regarding the use of new media. As more and more practices rely on technological mediation, yet we remain largely human and have practices with very human ambitions, we must consider whether valuable skilled practices may become lost or impaired as a result of the introduction of new technologies. Technologies are demanding in multiple ways, tied to both the affordances they are able to display for being successfully adopted into practice and the learning processes users must go through to become familiar with them. Although formal learning strategies are applied to counter such things, these are in the minority, and it remains primarily through practice-based learning that staff become familiar with the technological tools.

Lastly, the other major impact on responses to changing technologies in practice stems from the collaborative and collective nature of organizational life. During the unfolding actions of practice, people need to work together, negotiating with technologies on the one hand and interacting with colleagues and wider organizational concerns on the other. Technologies become a part of the organizational fabric of schools and hospitals through which collectively held notions and ways of doing emerge. This does not mean that we can easily separate the collective aspect form the individual engagement. Just as with the themes outlined earlier, they become entangled in complex confrontations and negotiations to arrive at adequate ways of aligning learning, actions, technologies, and their organization. Recollecting his own process of becoming familiar with a certain technology, a teacher explained,

> I'm self-taught on the computer, but here at work I needed more help and have read manuals by going on YouTube. We only went on a brief introductory course about how to use the interactive whiteboards (we did get that at least), but it was enough so that we could move on because we could talk to each other about it. "Hey, I have a problem here—do you know how to deal with that?" This was especially about technicalities . . . "my computer can't access the Internet . . . how can that be?" and then you learn three, four, or five possibilities, which may be wrong, but you try them. So it is going back and forth between trying what is possible, talking to others, and in the end a bit of formal education. But in this workplace we try to teach each other.

Through these practice-based learning experiences, the development of expertise in handling technologies relies on being able to understand the needs and motivations of others, or what could be termed "relational agencies" (Edwards, 2010). Such negotiations are necessary as technologies become meaningful phenomena linking tangible tools with thoughts, actions, and culture. Associated with the everyday routines of our local life-worlds, they help define our relationships and generate opportunities (Kim &Roth, 2008). Being not just helpful but mutually constituted in practice, they "bite back" and have "unintended consequences" (Tenner, 1996) that need addressing through collective understanding. Simply put, practitioners need to learn from each other how to successfully handle and communicate unpredictable consequences within and across disciplinary boundaries.

Edwards's notion of "relational agency" is pertinent here, involving a capacity for working together and giving support across boundaries. A nurse may learn about the electronic patient system from a doctor or a secretary. A schoolteacher may learn about the new "interactive whiteboard" from a friend who happens to be an IT consultant, and teachers with different backgrounds may learn from each other. We can see that a situated technological literacy involves forms of practice-based learning that include "relational agency." This isn't simply a capacity to work with technology or with colleagues but "the capacity to 'know how to know who' (can collaborate)" (Edwards, 2010, p. 31).

Seen in such terms, the reconfiguration of technology becomes interwoven between local individual and collective concerns and with the ever-widening aspects of organization and even beyond. There have been, in relative terms, huge investments in schools, for example, aimed at providing teachers with opportunities for creating multimedia presentations through the deployment of "interactive whiteboards." Presented to the schools as progressive and innovative, their repeated breakdown because of poor network connections results in them being considered as unreliable. In response, teachers reconfigure their approach to class preparation by ensuring they also have a contingency "plan B," adopting physical media such as books and printouts. The motivation for learning from each other dwindles and the opportunities for using the "interactive whiteboards" are inadequately explored, with the result that expensive and sophisticated technologies become adopted simply as a substitution for the blackboards they replaced.

In an attempt to draw together what we have seen as the primary strands or common experiences of sociomaterial interactions found from the study of the working professions of nursing and teaching, we can begin to propose how technological literacy is manifested within everyday working practices. Involving the ability to engage meaningfully with differing reconfigurations of technology within a context of professional culture that is inextricably linked with workplace learning, it can be approached from four interrelated dimensions. These relational dimensions provide a view of particular intersections of the mutual fashioning of technology and human activity. Two of these can be seen as disposed toward an initial technological consequence and the other two toward the social, although they all depend on the reciprocation between technological and social influence. These can be simply summarized as involving technologically biased constituents of "technological change" and "technological uncertainty" and socially biased constituents of "reflexivity" and "organization."

"Technological change" here refers to the learning consequences of deploying and implementing new technologies within the workplace. This involves becoming familiar with and adjusting to ways of doing things together with the affordances that technologies and their successive alterations provide toward the aims of practice. These are emergent technopractice realizations that demand an understanding and experience of the technology from the aspect of its direct use and its interrelations with the objectives for which it is adopted. This is the learning "how to use" dimension.

"Technological uncertainty" is the field of reconfiguration that results from unintended and unexpected technological consequences. This involves the indeterminate mismatch between technology and practice exemplified by the idea of "the breakdown" that prevents the unhampered completion of procedures and plans. Learning how to contend with the resulting uncertainty either in a response or analysis "in the moment" or through a broader understanding of technology is characteristic of this aspect. This is the learning "through events" dimension.

"Reflexivity" is the aspect of technological literacy able to situate technological demands within the social and human aspects of practice. It is where technologies can be understood as operating through ontologies and systems of logic that are different from those found elsewhere and that, without the reflexive ability to see beyond purely technological concern, can effectually "blind" the user from its consequences on a human level. This is the learning "between influence" dimension

"Organization" refers to reconfigurations that are situated within organizational concerns and those relating to the profession. These involve a metaperspective that goes beyond the immediate concerns of practice to incorporate collective consequences such as structuring, management, and policies that influence the longer-term impact of acting together with technology. This is the learning "together" dimension.

Technological literacy is the capacity for learning from everyday entanglements within the constant reconfigurations of both practice and technology without losing sight of the motive for the practice itself. A technologically literate practitioner therefore acts reflectively and actively among the continually shifting ecologies of practice (Wallace, 2010). Similarly, being passive to such changing conditions, one risks losing sight of the fundamental aims of practice and the learning through which this is maintained. This view of the interrelated intersections of technology and activity offers a particular perspective of the engagement in negotiating and learning together across the organization. It follows, then, that the capability of collectively keeping professional motives in mind during practice-based learning amid the shifting reconfigurations of practice and technology, astride the four dimensions of influence, defines the technologically literate person.

This attempt to free technological literacy from the confines of purely educational contexts has enabled it to be considered alongside the complexities of situated ways of acting. As such, this has reflected back the multiple dimensions of technologies' influence on our everyday lives, not least those found within education itself. The aspiration has been to provide a framework through which vocational and practice-based learning can view the inevitable intersections with progressive technological influence on professional life. This is perhaps a step toward empowering practitioners to develop relevant incentives (see Dow, 2006) for learning inpractice as well as about practice and to find ways to take charge of the opportunities technologies offer for the meaningful development of practice.

REFERENCES

Barnard, A. (2006). Technology, skill development and empowerment in nursing. In J. Daly, S. Speedy, & D. Jackson (Eds.), *Contexts of nursing* (pp. 199–212). Marrickville, Australia: Churchill Livingstone.

Billett, S. (2009, November). Realising the educational worth of integrating work experiences in higher education. *Studies in Higher Education, 34*(7), 827–843.

Brown, J.S., & Duguid, P. (1991). Organizational learning and communities of practice: Toward a unified view of working, learning, and innovation. *Organization Science, 2,* 40–57.

Carlile, P. (1997). *Transforming knowledge in product development: Making knowledge manifest through boundary objects.* (Unpublished dissertation). University of Michigan, Ann Arbor, MI.

De Vries, M. J. (2006). Technological knowledge and artifacts: An analytical view. In J. R. Dakers (Ed.), *Defining technological literacy: Towards an epistemological framework* (pp. 5–17). New York, NY: Palgrave Macmillan.

Dourish, P. (2006). Implications for design. In R. Grinter, T. Rodden, P. Aoki, E. Cutrell, R. Jeffries, & G. Olson (Eds.). *ACM Conference on human factors in computing systems*. Montreal, Canada: ACM Press.

Dow, W. (2006). Implicit theories: Their impact on technology. In J. R. Dakers (Ed.), *Defining technological literacy: Towards an epistemological framework* (pp. 239–250). New York, NY: Palgrave Macmillan.

Dugger, W.E., Jr. (2001). Standards for technological literacy. *Phi Delta Kappan, 82*, 513–517.

Edgerton, D. (2007). *The shock of the old: Technology and global history since 1900*. New York, NY: Oxford University Press.

Edwards, A. (2010). *Being an expert professional practitioner: The relational turn in expertise*. The Netherlands: Springer.

Feenberg, A. (2006). What is philosophy of technology? In J. R. Dakers (Ed.), *Defining technological literacy: Towards an epistemological framework* (pp. 5–17). New York, NY: Palgrave Macmillan.

Garmire, E., & Pearson, G. (Eds.). (2006). *Tech tally: Approaches to assessing technological literacy*. Washington, DC: National Academy Press.

Holland, D., Lachicotte, W. Jr., Skinner, D., & Cain, C. (Eds.). (1998). *Identity and agency in cultural worlds*. Cambridge, MA: Harvard University Press.

Ingerman, Å., & Collier-Reed, B. I. (2011). Technological literacy reconsidered: A model for enactment. *International Journal of Technology and Design Education, 21*, 137–148.

Kahn, R., & Kellner, D. (2006). Reconstructing technoliteracy: A multiple literacies approach. In J. R. Dakers (Ed.), *Defining technological literacy: Towards an epistemological framework* (pp. 254–273). New York, NY: Palgrave Macmillan.

Keirl, S. (2006). Ethical technological literacy as democratic curriculum keystone. In J. R. Dakers (Ed.), *Defining technological literacy: Towards an epistemological framework* (pp. 81–102). New York, NY: Palgrave Macmillan.

Kim, M., & Roth, W.-M. (2008). Envisioning technological literacy in science education: Building sustainable human-technology-lifeworld relationships. *Journal of Educational Thought, 42*(2), 185–206.

Lave, J., & Wenger, E. (1991). *Situated learning: Legitimate peripheral participation*. Cambridge, UK: Cambridge University Press.

Liddament, T. (1994). Technological literacy: The construction of meaning. *Design Studies, 15*(2), 189–213.

Michael, M. (2006). How to understand mundane technology: New ways of thinking about human-technology relations. In J. R. Dakers (Ed.), *Defining technological literacy: Towards an epistemological framework* (pp. 49–17). New York, NY: Palgrave Macmillan.

Orr, J. E. (1996). *Talking about machines: An ethnography of a modern job*. Ithaca, NY: Cornell University Press.

Star, S. L. (1989). The structure of ill-structured solutions: Boundary objects and heterogeneous distributed problem solving. In M. Huhns & L. Gasser (Eds.), *Readings in distributed artificial intelligence* (pp. 251–273). Menlo Park, CA: Morgan Kaufman.

Suchman, L. A. (2007). *Human-Machine Reconfigurations: Plans and situated actions* (2nd ed.). Cambridge, UK: Cambridge University Press.

Tafdrup, O., & Hasse, C. (2012). Praksislæring af teknologiske artefakter (Practice-learning of technological artefacts). In C. Hasse & K. Søendergaard (Eds.), *Teknologiforståelse* (Understanding technology) (chap. 10). Aarhus: Aarhus University.

Tenner, E. (1996). *Why things bite back: Technology and the revenge of unintended consequences.* New York, NY: Knopf.

Waks, L. J. (2006). Rethinking technological literacy for the global network. In J. R. Dakers (Ed.), *Defining technological literacy: Towards an epistemological framework* (pp. 275–273). New York, NY: Palgrave Macmillan.

Wallace, J. (2010). *Different matters of invention: Design work as the transformation of dissimilar design artefacts.* (Unpublished dissertation). The Danish School of Education, Aarhus University.

Wallace, J. (2012). Rekonfigurering af teknologier i sygeplejepraksis: Fra indført til foretrukket. (Reconfiguring technologies in nursing practice: From the imposed to the preferred). In C. Hasse & K. Søendergaard (Eds.), *Teknologiforståelse* (Understanding technology): På lærer og sygeplejerske arbejdspladser (chap. 9). Aarhus: Aarhus University.

Yawson, R. M. (2010). An epistemological framework for nanoscience and nano-technology literacy. *International Journal of Technology and Design Education.* doi:10.1007/s10798-010-9145-1.

GENETIC LITERACY

SCIENTIFIC INPUT AS A PRECONDITION FOR PERSONAL JUDGMENT?

Silja Samerski

The nonsquashy tomato from the genetic engineering laboratory is called a "GM," or genetically modified, tomato in popular parlance, and the manipulated food of the genetic engineering industry is called "GM food." Surveys show that many people assume that these "genetically modified foods" contain genes, while tomatoes from their own gardens do not.[1] "The majority do not know that they are always chewing genes," comments the newspaper *FAZ* (Müller-Jung, 2006). And plant geneticist Hans-Jörg Jacobsen questions whether such ignorant citizens can even participate in a democracy, asking "how our democratic society can make decisions on this kind of meager and uninformed basis. A proper public discourse is called into question and the floodgates are open for one-sided ideologically based agreements" (Jacobsen, 2001, my translation).

This lament over an unenlightened population and the necessity of disseminating "proper" information is widespread (Kerr, 2003; The Genetic Literacy Project, 2012).[2] In the genetic age, the proper governance of science, politics, and industry requires citizens who can participate in biopolitical debates and make "informed decisions." Government agencies and professionals wish to remedy this bemoaned ignorance by promoting genetic literacy (Jennings, 2004). The "life sciences," and genome research in particular, predicts the German Federal Ministry of Education and Research,[3] will have "far-reaching effects" on "our entire social life" (BMBF, my translation). To prepare the population for these pervasive upheavals, the ministry is funding genetic education in the belief that "secondary research" in the social sciences and on the "discourse between science and society" should contribute toward producing a "well-informed public" (BMBF, 2006, my translation). The goal, according to the ministry, is "to be able to base decision-making on comprehensible facts and rational arguments" (BMBF, 2013).

Industry is likewise committed to promoting genetic literacy. Bayer, Schering, and Roche all want rational, decision-making citizens and informed consumers. "We know that the greatest developments are worth nothing when people do not understand them and thus are not prepared to accept them," explained the chairman of the Association of Chemical Industry (VCI) in North Rhine in 2001, and he sent off a mobile genetic laboratory to promote proper understanding among citizens (Minwegen, 2003, my translation). A number of institutions have thus taken on the mission of promoting the genetic literacy of the population. Science centers, websites, discourse projects, citizen conferences, life science learning laboratories, physician-patient informed-consent discussions, and genetic counseling clinics all try to turn genetic illiterates into genetic citizens (Heath, Rapp, & Taussig, 2004). As different as the educational program and the public may be, they share one goal in common: they strive to mobilize citizens, by means of professional instruction, to engage with "genes" and "risks" and become qualified to participate in informed debates and make informed decisions. Citizens should know that tomatoes and people carry genes, and they should base their decisions on these facts.

This attempt to facilitate "informed decisions" by promoting genetic literacy is the theme of this chapter. I want to challenge the assumption that genetic education empowers citizens to make an independent personal judgment. On the basis of my empirical research on "genetic alphabetization" in Germany (Samerski, 2010), I analyze the hidden curriculum of genetic education. What is being asked of citizens when autonomy and responsibility requires having experts update your own powers of judgment? Doesn't the alleged "proper" information already provide the bottom line—namely, the framework and basis for deliberations? In what form of thinking are citizens being initiated when they are no longer supposed to be moved by experience and tangible realities but rather by scientific constructs? These questions loom large, especially since the notion of the "gene" as a definable, controllable, and causative entity is scientifically antiquated (Keller, 2002). If the concept of the gene is obsolete in research, what do people learn when their genetic literacy is improved?

Drawing on three instances from my participant observations at public events and genetic counseling sessions, I will argue that the promotion of genetic literacy disables citizens' common sense by instilling in them the scientifically buttressed worldview of geneticists. First, at a public congress aiming at fostering public debate on gene technology, the experts related not to common sense (Arendt, 1958) but to scientific terms and expert opinions. Thus they widened the gap between science and the public and rendered the audience speechless. As my second example, I take the public lecture by a leading German geneticist to show that the truth claims concerning genes are part of the creation of an aura of indisputability around genetic expertise. The geneticist explained how genes "really" influence human beings, thereby claiming unassailable authority over people's bodies and their very beings. In my third example, genetic counseling, the genetic lessons are claimed to have direct personal relevance to a specific individual. Clients are instructed not only about genes and humankind in general but about themselves in particular. Thus they are invited to take genetic constructs as revelations of their very being, their health and disease, their mind and body, their past and future. Finally, I conclude that campaigns for genetic literacy propagate submission under the defining power of genetic expertise and make scientific input a precondition for responsibility and autonomy.

RENDERING THE PUBLIC SPEECHLESS

"Good Genes—Bad Genes?" was the title of a national congress for civic education held in Bremen (Germany) in early September 2003. The organizers, the Federal Agency for Civic Education[4] and its Bremen branch saw the dawning of an era of genetic technology and considered it their duty to prepare the general public. The congressional announcement asserts that "the biosciences and biotechnologies are currently learning to understand, control, even improve fundamental life processes at a tremendous speed" (BpB, 2003, my translation). Thus the sponsors of the congress hoped to enable citizens "to be an active influence on the decision-making process" (BpB, 2003, my translation). To this end, they invited to Bremen three dozen high-ranking experts from around the world who discussed, among other issues, the ethical defensibility of stem-cell research, the church- and state-implemented eugenics program in Cyprus, and the responsibility of parents for their children's genetic makeup.

For their part, the experts were clear about the cause of the lack of democratic participation: their diagnosis is that people are hopelessly backward where genetics is concerned. When it comes to DNA, heredity, and genetic testing, most people are ignorant and disempowered. They have opinions, but for the organizers and the experts, the vox populi is too unqualified. "Perceptions," complained the director of the Bremen agency, but not "knowledge" currently shape the attitudes of citizens toward gene technology. And the director of Bremen's Center for Human Genetics attributed reservations about his field to "misinformation": unrealistic hopes, he declared succinctly, lead to unrealistic fears. Therefore he prescribed counseling and education for his fellow citizens to enable them to deal rationally with genetics.

Also, industry insisted on the need for an educated citizen able to participate in "democratic biopolitics." What they hope to gain from this was made very clear by the pharmaceutical representative from Roche toward the end of his presentation: society, not industry, bears the "responsibility" for gene technology, he explained. "Society must decide what it wants to do with gene technology," he demanded. In his presentation, it quickly became clear, however, that industry does not intend to simply subject itself to the will of the people. So far, he complained, people have lacked the "knowledge" and the "assessment opportunities." Society, he made clear, must first be prepared for this new challenge. He literally said, "We must help society understand . . . We must explain to society how it should understand and how it must decide."

The congress did not achieve the goal set by the Federal Agency for Civic Education. On the contrary, for the most part, a public debate failed to materialize. No discussion took place among the citizens attending the congress. Only a few members of the audience spoke after the presentations. Apparently, most listeners were rendered speechless. Although the presentations from experts were intended to stimulate discussion, they were not delivered in a manner that would be broadly comprehensible to laypeople but instead were peppered with specialized professional jargon. The speakers talked about "zygotes" before and after "nuclear fusion," "chromosome aberrations," and "genes for" various unknown diseases, as well as "disease probabilities," "genetic dispositions," and "risk carriers." It was assumed that the starting point of a democratic discussion is not common sense but the scientific terms used in the discourse of experts. The conference discussed not Bremen citizens' experiences, fears, and desires but scientific laboratory constructs and bioethical problems. Many presentations only touched on people and their experiences when the geneticists resorted to commiserating tales of woe to plead for research money and deregulation.

The Bremen congress is an impressive example of the attempt to educate citizens so they can attain greater self-determination in genetics-relevant issues. However, instead of dismantling barriers to democratic participation, the congress set up new ones. The organizers convened experts who explained that their professional knowledge is an essential precondition for a democratic discussion and that the entire population is consequently in need of counseling. Only those who are instructed by geneticists and bioethicists should have a voice about gene technology—in regard not to scientific but to social issues. The proposed topic of the congress was not the scientific perspective on gene function and DNA structure; rather, it was supposed to be the effects of a new technology on society. Precisely here is where the speakers denied their fellow citizens their capability and judgment. And in doing so, they undermined the basis for democratic discussion.

THE GENETIC MUTATION OF THE LISTENERS

When geneticists convey their expertise to the public, they claim to have authoritative knowledge not only about genes but also about their fellow citizens. Almost inevitably, by talking about "genes" in a popularized way, they reframe their listeners and interlocutors (Duden & Samerski, 2007).

The director of the Hanover Institute for Human Genetics has taken it upon himself to draw the public's attention to distorted notions about genes and to appropriately convey "the role of genes in human life" (p. 172). Right at the beginning of his speech, he points out to his listeners that they are the objects of his expertise: he will be talking about the "human species to which we all belong" (p. 172), he clarifies. Within this species, there are only insignificant genetic differences, he goes on to explain. These sentences set the framework for the seminar. First he appropriates everyone in attendance, whether they like it or not, into an inclusive, global, and inescapable "we all" as the "human species" (p. 173). Then he appoints himself as an expert about this biological "we" by declaring that all people are gene carriers. The geneticist talks to his listeners not as peers but as members of a biological species about which he possesses scientifically validated knowledge. Although the people in the audience are the targets of his explanations, they are also the objects of his specialized knowledge. Objections or protestations of common sense have no place here. As a geneticist, he only knows about gene carriers: in other words, about the people in the audience. A common ground for dialogue therefore does not exist.

As a psychogeneticist, he is searching for "behavior-guiding genes" (p. 173). To this end, he is researching the mating behavior of the female rhesus monkey. Hence he feels empowered to explain the causes of marital infidelity to his listeners—mostly women. As he has already reframed "being human" as belonging to the biological species *Homo sapiens,* he can now talk about the limbic system, copulating female monkeys, and marital fidelity in a single breath. He knows—only a few genes, after all, separate female monkeys from women—that serotonin levels contribute to the occurrence of extra-marital escapades by the female sex. As a champion of genes that do not determine but merely dispose, however, he does not want to simply excuse the unfaithful. "Genes," he explains, are "cross-linked information carriers" that "sometimes crash" and receive external "commands" (p. 175).Thus human beings are not victims of their genes. They can learn to live with them. The prerequisite for this, however, is counseling by geneticists. Those who do not want their genes to give them the runaround must become aware of and be informed about their genetic inheritance and then actively choose their behavior. Instead of giving in to the infidelity gene, suggests the expert casually, one could eat chocolate instead. Chocolate contains a serotonin precursor.

Human geneticists reframe people into two-legged gene carriers and make it clear to them that they need genetic education. As a bundle of DNA, mutations, and hidden information units, they can no longer know themselves. Those who wish to be "autonomous" must go to a geneticist to find out what their "self" actually is. In the age of genetics, *autonomous* no longer means being without supervision; rather, it presupposes having been taught about oneself by a genetic expert.

THE POPULAR GENE

Those who learn about genetics by attending congresses or seminars are free afterward to go home and return to their everyday lives. There is, however, a form of genetic education that has direct personal relevance and that requires citizens to act: genetic counseling. Genetic counseling is a professional service that prepares people to make a concrete decision—typically an "informed decision" about whether to undergo genetic testing or not. The explicit goal of the session is "medically competent, individual guidance in decision making" (Kommission für Öffentlichkeitsarbeit und ethische Fragen der Gesellschaft für Humangenetik e.V., 1996, p. 129). Thus genetic counseling is a paradigmatic example of a genetics educational event. Typically, a session lasts one to two hours. A genetic expert explains DNA and chromosome structure, heredity rules, disease statistics, and genetic testing options. If the geneticist is sitting across from a pregnant woman who needs to decide whether or not to undergo prenatal testing, he emphasizes the cellular processes involved in fertilization, statistics of birth defects, pregnancy risks, and prenatal testing options. On the other hand, if sitting in front of him are men and women whose families have had multiple occurrences of breast or colon cancer, then the topics of genetic mutations, cancer statistics, and early detection usually take precedence (Samerski, 2005).

The goal of the lessons on "genes" is to teach the client about himself. To do so, the geneticist first points out to her client that, like all people, he is a gene carrier. She explains genes, DNA, chromosomes, genetic information, and mutations and asks him to see himself as the product of these invisible gene worlds. She speculates about the specific gene errors and disease genes that her client may have. In these instructions, the complex and largely ambiguous relationship between genotype and phenotype (Lewontin, 2004) jells into an active linear causation. Oftentimes, counselors speak of "genetic defects" and "gene errors," invoking everyday connotations in order to explain a statistical correlation. A "genetic defect" seems to cause a disease like a cylinder defect that is the cause of an engine breakdown.

Geneticists formulate numerous sentences in which the gene is the subject of a verb indicating a causative activity. A woman in her forties learns, for instance, that she may have a gene mutation: "And this mutation can specifically result in (–) specific diseases; in other words, it can trigger them." Such formulations where genes are the subjects of active verbs are very prevalent. Like the "gene error," they convey meanings from everyday language to the abstract sphere of markers, statistical associations, and probabilities. In the process, the gene takes on animist traits: it becomes the mover that initiates and acts on its own.

The reification of "genes" and "gene errors" is reinforced by the notion that something is being stored there: "genetic information." Gene mutations, clients are taught, are mutations in the genetic code—that is, in the person's blueprint. A young pregnant woman, for example, is taught that genetic information contains information about the entire person: "Every cell has all the information about what constitutes the person . . . This means that when I remove one cell from someone's body, I have representative genetic information about the entire person" (Samerski, 2002, p. 157). Therefore, genes have intangible content; they contain information, instructions, or "administrative directives" (Samerski, 2002, p. 156).

CONCLUSION

Educating people about their genes and preparing them for informed decision making, geneticists turn the "gene" into a tangible and forceful reality. In scientific research, such a real, objective gene does not exist. Here, "gene" has a precise meaning only within very specific experimental practices: in other words, when researchers using the same methods are working on the same problem (Keller, 2002). Science philosopher Philip Kitcher sums up the status of the term *gene* in genetics: "A gene is anything a competent biologist chooses to call a gene" (Kitcher, 1992, p. 131). Therefore, the reified and all-explaining gene conveyed during educational events cannot be discounted as the popular science distortion of a real, objective gene in the laboratory. Instead, it has to be understood as the cornerstone of a powerful ideology. Today the genetics project still lives on the "genes in our heads" ideology (Duden, 2002): from the belief that there are genes that determine the phenotype and regulate the organism, that genetics is encoding the mystery of life and ultimately will contribute toward creating a better world. These convictions have spurred genetic research and dictated research questions, methods, and research findings. The reified gene therefore not only is the result of popular science communications but also is the ideological complement of the scientific gene industry. And this ideological complement, the "gene in the head" of geneticists, appears as soon as they promote genetic literacy and convey their expertise to their fellow citizens. As soon as geneticists attempt to depict their expertise as relevant to everyday life, they inevitably fall back on their own "genes in their heads": they ascribe to their clients those genes that correspond to popular scientific notions, their genetically based worldview.

Genetic literacy is praised as the basis of an active, self-empowered way of life. According to science, politics, and industry, only citizens well versed in genetics can be responsible citizens capable of taking destiny into their own hands. As my analysis has shown, however, the promotion of genetic literacy almost inevitably subverts a freedom: the freedom to know and decide for oneself—without professional guidance. The need to consume professional instructions in genetics in order to be considered mature for autonomous and responsible decisions perverts the once emancipatory call not to be patronized by authorities but to be guided by one's own intellect. Immanuel Kant's "Sapere aude!" ("dare to know") as the battle cry of the Enlightenment is subverted in the duty to make an "informed choice" guided by scientific input.

NOTES

1. In the Eurobarometer, this question has been asked for several years, see lastly Eurobarometer 2006, 57.
2. For a critique of the underlying "deficit model," see Hilgartner (1990).
3. Bundesministerium für Bildung und Forschung (BMBF).
4. Bundeszentrale für politische Bildung (BpB) and Landeszentrale für politische Bildung.

References

Arendt, Hannah. (1958). *The human condition*. Chicago, IL: University of Chicago Press.

BMBF. (2006). Bekanntmachung. Retrieved October 8, 2013, from http://www
.gesundheitsforschung-bmbf.de/de/1276.php.

BMBF. (2010). Lebenswissenschaften (Life sciences). Retrieved March 27, 2010,
from http://www.bmbf.de/de/1237.php.

BMBF. (2013). Life sciences. Retrieved October 8, 2013, from http://www.bmbf
.de/en/1237.php.

BpB. (2003, September). Veranstaltungsdokumentation: Gute Gene–schlechte Gene?
Gentechnik, Genforschung und Consumer Genetics. Retrieved October 8, 2013,
from http://www.bpb.de/veranstaltungen/dokumentation/129713/gute-gene
-schlechte-gene.

Duden, Barbara. (2002). *Die Gene im Kopf—der Fötus im Bauch*. Hanover, Germany:
Offizin.

Duden, Barbara, & Samerski, Silja. (2007). "Pop-genes": An investigation of "the
gene" in popular parlance. In R.V. Burri & J. Dumit (Eds.), *Biomedicine as culture:
Instrumental practices, technoscientific knowledge, and new modes of life* (pp. 167–
189). New York, NY: Routledge.

Genetic Literacy Project. (2012). Genetic literacy project: Where science trumps ideology. Retrieved October 7, 2013, from http://www.geneticliteracyproject.org.

Heath, Deborah, Rapp, Rayna, & Taussig, Karen-Sue. (2004). Genetic citizenship. In
Frank A. Nugent & Joan Vincent (Eds.), *A companion to the anthropology of politics*
(pp. 152–167). Malden, MA: Blackwell.

Hilgartner, Stephen. (1990). The dominant view of popularization. Conceptual
problems, political uses. *Social Studies of Science, 20*, 519–539.

Jacobsen, Hans-Jörg. (2001). Angemerkt: Prof. Dr. Hans-Jörg Jacobsen. Retrieved
October 8, 2013, from http://www.loccum.de/folo/angemerkt/jacobsen.html.

Jennings, Bruce. (2004). Genetic literacy and citizenship: Possibilities for deliberative
democratic policymaking in science and medicine. *The Good Society, 13*(1): 38–44.

Keller, Evelyn Fox. (2002). *The century of the gene*. Cambridge, MA: Harvard University Press.

Kerr, Anne. (2003). Genetics and citizenship. *Society, 40*(6): 44–50.

Kitcher, Philip. (1992). Genes. In Evelyn Fox Keller & Elisabeth A. Lloyd (Eds.), *Keywords
in evolutionary biology* (pp. 128–131). Cambridge, MA: Harvard University Press.

Kommission für Öffentlichkeitsarbeit und ethische Fragen der Gesellschaft für
Humangenetik e.V. (1996). Positionspapier. *Medizinische Genetik, 8*, 125–131.

Lewontin, Richard. (2004). The genotype/phenotype distinction. *The Stanford encyclopedia of philosophy*. Retrieved October 7, 2013, from, http://plato.stanford.edu/
entries/genotype-phenotype.

Minwegen, Norbert. (2003). Biotechnologie und Gentechnik zum Anfassen. Retrieved
October 7, 2013, from, http://www.projektwerkstatt.de/gen/filz/quellen/
kap6konzerne_ende29_32.pdf.

Müller-Jung, Joachim. (2006, September 9). Am Gen erstickt. *FAZ*: N1.

Samerski, Silja. (2002). *Die verrechnete Hoffnung. Von der selbstbestimmten Entsche-
idung durch genetische Beratung*. Münster, Germany: Westfälisches Dampfboot.

Samerski, Silja. (2005). Genetic counseling. In Carl Mitcham (Ed.), *Encyclopedia of
science, technology and ethics*. New York, NY: MacMillan.

Samerski, Silja. (2010). *Die Entscheidungsfalle. Wie genetische Aufklärung die Gesellschaft entmündigt*. Münster, Germany: Wissenschaftl buchgesell.

ABOUT THE CONTRIBUTORS

Andoni Alonso (Vitoria, Spain), received his PhD in philosophy in 1995 from the Universidad del País Vasco and was a research fellow at Penn State University (1996–98) and Nevada University at Reno (2004). He served as an assistant professor at the Universidad de Extremadura and is currently an associate professor at Universidad Complutense de Madrid. His research field is science, technology and society, and computer philosophy, and his interests include philosophy of technology, cyberculture, amateur and 2.0 science, net artivism, and political movements through the net. His publications include *Todos Sabios: Ciencia Ciudadana y Conocimiento Expandido* (Madrid: Cátedra, 2013; with Antonio Lafuente and Joaquín Rodríguez); *Diasporas in the New Media Age* (Reno: Nevada University Press, 2009; with Pedro Oiarzabal), *La Quinta Columna Digital* (Barcelona: Gedisa, 2004; awarded the Epson Prize on Techno-Ethics), and *La Nueva Ciudad de Dios* (Siruela: Madrid, 2002).

David Barlex is an acknowledged leader in design and technology education, curriculum design, and curriculum materials development. He taught science and technology in comprehensive schools for 15 years before becoming a teacher educator. He directed the Nuffield Design and Technology Project, which produced an extensive range of curriculum materials widely used in primary and secondary schools in the United Kingdom and has been emulated abroad—Russia, Sweden, Canada, South Africa, Australia, and New Zealand. He was the educational manager of Young Foresight, an initiative that has developed approaches to teaching and learning that enhance students' ability to respond creatively to design and technology activities. In 2002, he won the DATA Outstanding Contribution to Design and Technology Education award. Barlex's research activity stems from his conviction that there should be a dynamic and synergic relationship between curriculum development and academic research. His research interests include pedagogy that develops design ability and creativity and the professional development of teachers. He has presented work regularly at international conferences and published in the *Journal of Design and Technology* and the *International Journal of Technology and Design Education Research*. Most recently, he has acted as the STEM (science, technology, engineering, and mathematics) consultant and a curriculum adviser for the Digital Design and Technology project to the Design and Technology Association.

John R. Dakers is affiliated with the Technology University of Delft in the Netherlands (TUDelft). His research interests include technology education, the philosophy of technology, technological literacy, gender issues related to technology, and the concept of communities of learners. He has written and been published on these subjects in journals and in books, has acted as guest editor for several journals, and has edited several books, including this second edition.

He acted as conference director for two major international conferences on "Pupils Attitudes Towards Technology" (PATT), both held in Glasgow in 2003 and 2006. He has been invited to deliver keynote presentations around the world, including in the United States, France, the United Kingdom, Finland, and Sweden. His first book, *Defining Technological Literacy*, published in 2006, is about to be published as a significantly revised second edition. He is currently working on several books, including one on the philosophy of technological education and a monograph on technological literacy.

Leo Elshof is an associate professor at Acadia University in Wolfville, Nova Scotia, Canada. He has worked in the fields of science, engineering, and sustainability education for thirty years. He holds degrees from the University of Waterloo, McMaster University, and the University of Western Ontario and a doctorate from the University of Toronto. At Acadia, he teaches courses in environmental and sustainability studies and preservice and graduate education programs in the areas of science and environmental education, critical media literacy, and environmental technologies. Elshof has published articles in a number of international journals, authored numerous book chapters, and presented at more than forty national and international academic conferences on the themes of sustainability in technological and science education. His current research interests involve eco-technological literacy, climate change education, critical media literacy, and environmental communications.

Cathrine Hasse is a professor and research program leader of Future Technologies, Culture and Learning at the department of Education at Aarhus University. She has expertise in studying learning, technology, and culture in organizations with a special focus on workplaces such as universities and colleges, vocational education, schools, and hospitals.

Her training as an anthropologist is specialized in cultural anthropology, and she has extensive knowledge of learning theory—especially in cultural psychology—and is versed in postphenomenological philosophy of technology as well as feminist science studies. Her main area of study is cultural learning processes in science and technological studies. Her many years of academic work have served to inform university departments and other workplaces such as schools, nursing homes, and hospitals. Within the field of anthropology, she has specialized in the relatively narrow field of anthropology and learning and ethnographic methodology connected to learning.

Her main contribution is a new theory of cultural learning processes that underlines the relation among moving in physical space, materiality, and collective meaning-making processes. These insights have been contextualized by her coordination of a number of projects financed by the Danish Research Council, as well as an EU project. In these projects, she has explored the relation between learning in cultural organizations and materiality and play with technology and has developed several specific concepts tied to the notion of cultural learning processes: "the relational zone of proximal development," "social designation," and the relation between agential knowing and *sprezzatura*. In her capacity as evaluator of education, she has been involved in the evaluation of teaching the subject of pedagogy in Danish professional university colleges and especially teachers and nurses education in, for example, the ongoing *Technucation* project (http://www.technucation.dk).

Mary Kirk is a professor and department chair of individualized, interdisciplinary, and lifelong learning at Metropolitan State University in Saint Paul, Minnesota, where she has taught since 2000. Kirk's work in feminist science studies includes her book *Gender and Information Technology: Moving Beyond Access to Co-Create Global Partnership* (2009); publications in a variety of journals such as *NWSA Journal, NWS Action, Feminist Teacher*, and the *Journal of Computing in Small Colleges*, and presentations at conferences such as the Grace Hopper Celebration of Women in Computing and the National Women's Studies Association. Kirk was a founding faculty member of the computing and software systems program at the University of Washington, Bothell (UWB), where she taught technical writing, women in computing, and ethics in computing as a lecturer while completing her PhD (2000) in women's studies/women in computing from Union Institute and University. The late Dr. Anita Borg served on Kirk's doctoral committee, which led to Kirk organizing and hosting the first Institute for Women and Technology (now named the Anita Borg Institute) workshop at UWB. Kirk also holds an MA (1996) in women's literature from the University of Illinois at Springfield. Prior to her academic career, Kirk was a technical writer at companies such as IBM, Schlumberger, and Microsoft. Kirk is currently working on her second book.

Carl Mitcham is a professor of liberal arts and international studies at the Colorado School of Mines. He holds visiting appointments at the Center for Science and Technology Policy Research, University of Colorado; European Graduate School; and the Consortium for Science, Policy, and Outcomes, Arizona State University. His publications include *Thinking through Technology: The Path between Engineering and Philosophy* (1994), *Humanitarian Engineering* (with David Munoz, 2010), *Ethics and Science: An Introduction* (with Adam Briggle, 2012), and the *Encyclopedia of Science, Technology, and Ethics* (4 vols., 2005; a second revised edition of is scheduled for 2014). A book on philosophy of engineering was recently published in Chinese and will eventually appear in English.

Stephen Petrina specializes in media and technology studies and science and technology studies (STS). He is a cultural historian with current interests in the philosophy of research, as opposed to methodology. A critical educator and professor (media and technology studies), he is in the Department of Curriculum and Pedagogy at the University of British Columbia.

Petrina has also published a popular textbook titled *Advanced Teaching Methods for the Technology Classroom* (2007) and a collection of essays on the critical theory of design and technology education. Recent articles appear in the *History of Education Quarterly* (e.g., "Preschools for Science," an award-winning article with Penney Clark and Mona Gleason), *History of Psychology, Technology and Culture*, and the *International Journal of Technology and Design Education*.

Silja Samerski is a biologist and sociologist. In 1996, she earned her diploma in the Department of Human Genetics, at the University of Tubingen, Germany with a thesis in population biology. While working on the genetic makeup of Madagascar monkeys, she became aware of the ambiguity of technical terminology once it migrates from the laboratory. Terms such as "mutation" or "genotype" have a precise denotation for biologists but are loaded with everyday meaning as part of ordinary conversations and become powerless to denote anything. She earned her PhD at the University of Bremen focusing on genetic counseling as a paradigm for the popularization of genetic concepts. Afterward, she collaborated with the historian Barbara Duden on the research project *Das Alltags-Gen* (The "pop-gene"), studying the meanings and connotations given to the word "gene" when it is used in everyday conversations. In her ongoing research, Samerski explores the genealogy of the patient as a decision maker and the occurrence of a "regime of decisions" as a new technique of social engineering. Since January 2013, she has been a research fellow at the research training group "Self-Making: Practices of Subjectivation" at the University of Oldenburg.

Jamie Wallace is a postdoctoral researcher in the Department of Education, Aarhus University. He also holds degrees in mechanical engineering and in art and design and has a master's degree in fine art. His general research interests involve material and visual artifacts and cultures of working practice. His interdisciplinary involvement in art, design, and technology is rooted in his own experiences of working within engineering design and the visual arts and as a coal miner. Intimate knowledge of these working practices continues to inspire and inform his research and writing. He gained his PhD in 2010 after studying the use of artefacts while working for a Danish industrial design agency designing equipment for the loading of aircraft. His study explored the interrelated roles played by physical and digital media in the design process. Concern was for how the engagement of materialities had consequences for embedded creativity, innovation, and design management. Most recently, he is involved with the Technucation research project funded by the Danish Council for Strategic Research, where his focus has been the situated relations between users and technology and the ways new technologies lead to a reconfiguration of the working practices of Danish nurses and teachers.

Technological literacy has been central to the study, providing insights into workplace learning in the face of disruptive technologies. His involvement here also includes the development of learning tools promoting technological literacy within vocational education. He recently published "Emergent Artefacts of Ethnography and Processual Engagements of Design" in *Design and Anthropology*, edited by Jared Donovan (Ashgate 2012).

Nan Wang has an MA in the philosophy of science and technology from the Dalian University of Technology and PhD in philosophy of science and technology from the Graduate University of the Chinese Academy of Sciences. She worked as a postdoctoral student from January 2012 to March 2013 funded by the Hennebach Program in the Humanities at the Colorado School of Mines. She is an assistant professor at the School of Humanities and Social Sciences at the University of the Chinese Academy of Sciences. Her research areas include the philosophy of technology, philosophy of engineering, and sociology of engineering. Her recent publications include "The Development of Railroads in America and China" in *Engineering, Development, and Philosophy: American, Chinese, and European Perspectives*, edited by Steen Hyldgaard Christensen et al. (2012), "What Is the Character of the Techno-Human Condition?" (coauthored with Wenjuan Yin) in *Technè* (Fall 2012), and "Philosophical Perspectives on Technology in Chinese Society" in *Technology in Society, 35,* no. 3 (2013).

Molly Watson is a 15-year-old student attending a Scottish secondary school. She has a passion for writing in various forms and is an active member of her local Creative Writing Society, through which she has had a piece of poetry published in a national anthology. She also has a keen interest in history and politics and likes to observe the amateur debates within her school (although as yet she is often too shy to participate). At the age of ten, Molly became ill with myalgic encephalomyelitis, and her health was badly affected for a few years. This is when she began to discover her love of writing, using pen and paper as an outlet and a confidant. Now she is recovered and aspires to a career in writing fiction or poetry.

P. John Williams, associate professor, is the director of the Technology, Environmental, Mathematics and Science Education Research Centre at the University of Waikato in New Zealand, where he teaches and supervises research students in technology education. Apart from New Zealand, he has worked and studied in a number of African and Indian Ocean countries and in Australia and the United States. He directed the nationally funded *Investigation into the Status of Technology Education in Australian Schools*. His current research interests include mentoring beginning teachers, Pedagogical Content Knowledge, and electronic assessment of performance. He regularly presents at international and national conferences, consults on technology education in a number of countries, and is a longstanding member of eight professional associations. He is the editor of the *Australasian Journal of*

Technology Education and is on the editorial board of four other professional journals. He has authored or contributed to more than two hundred publications, and in 2011 he was elected to the International Technology and Engineering Education Association's Academy of Fellows for prominence in the profession. He holds an adjunct professorial position at Edith Cowan University in Perth, Western Australia.

Index

Author

SUBJECT

Printed and bound by CPI Group (UK) Ltd, Croydon, CR0 4YY